Farm Irrigation: Planning & Management

Farm Irrigation: Planning & Management

NEIL SOUTHORN

INKATA PRESS

INKATA PRESS
A division of Butterworth-Heinemann Australia
Australia
Butterworth-Heinemann, 18 Salmon Street, Port Melbourne, 3207
Singapore
Butterworth-Heinemann Asia
United Kingdom
Butterworth-Heinemann Ltd, Oxford
USA
Butterworth-Heinemann, Newton

National Library of Australia Cataloguing-in-Publication entry

Southorn, Neil.
Irrigation and drainage.

Includes index.
ISBN 0 7506 8937 4.

1. Irrigation farming. 2. Drainage. I. Title. (Series :
Practical farming).

631.587

Typeset by Ian MacArthur, Hornsby Heights, NSW.
Printed in Singapore by Chung Printing.

Contents

Acknowledgments

Special thanks to Joanne for assisting with the manuscript, and the following organisations:

Netafim Australia Pty Ltd
Central Highlands Management
NSW Agriculture
Irricrop Technologies Pty Ltd
Sentek Pty Ltd
Electronic Irrigation Systems
James Hardie Irrigation Pty Ltd
Vinidex Tubemakers Pty Ltd

Introduction

Irrigation makes a major contribution to agricultural production by making a whole range of crops viable in an otherwise unreliable climate, and helping insure against drought. However irrigation does not automatically guarantee profit and acclaim. It is a high cost exercise, using water from increasingly limited supplies, and raises environmental concerns in the community. Many of the pressures facing some irrigators have been caused by a lack of understanding in the past of the best practices necessary for design, installation and management. This book attempts to summarise the key factors and processes in successful irrigation.

Chapters 5, 6 and 7 describe different irrigation methods. Existing irrigators may be looking to convert to a different method, or to modify their existing system to use less water or reduce labour costs. New entrants to the industry will need to decide the best method of irrigation for their situation. The book covers irrigation methods for larger scale agricultural and horticultural crops, and does not cover landscape irrigation or hydroponics.

The technology associated with the detailed design of irrigation systems is quite specialised and best left to experts. Irrigation systems work to relatively tight specifications. It would be unusual for landowners to design and install their own equipment or earthworks. Consequently, this book provides only an introduction to irrigation methods, to enable consideration of the best watering method for specific situations, sufficient for feasibility planning.

Sometimes, that choice is pre-determined. For example, irrigation of orchard or plantation rows in undulating to steep country can only be effectively achieved by microirrigation. Where the site is relatively flat, soil types suitable, and environmental hazards minimal, the choice of irrigation methods may be determined more by economic factors than technical ones.

Regardless of the method of watering, there are some common principles that underlie irrigation planning and management, including the need for efficient and uniform water application, accurate matching of irrigation amounts to crop water requirements, consideration of growing conditions in the root zone, and the ability to get enough water during critical peak periods.

Irrigation is concerned with providing the optimum soil moisture conditions for plant growth. So too is drainage, in that too much water in the soil will retard growth. Many of the concepts of irrigation apply to drainage considerations, and Chapter 9 discusses this topic as well.

An understanding of the way water is held in the soil, and the transpiration process by which plants consume soil water, is an essential pre-requisite of successful irrigation. Chapters 2 and 3 cover these processes in detail, but the key terms and concepts are explained below.

Readers need to be clear on how they might apply information from this book. In particular, there is a difference in the way certain decisions might be made for design of an irrigation system, compared to its management. For example, the irrigation designer needs to select the capacity of the equipment (the size of pipes and channels, and the performance of pumps) based on an estimate of the maximum likely demand for water for the particular crop, soil type and weather pattern for the locality. Similar information is monitored on a daily basis to make management decisions regarding irrigation scheduling, but it is applied differently.

Water balance of the root zone

Figure I.1 shows a plant growing in soil. The plant extends its root system a certain depth into the soil, referred to as the "root zone". For optimum plant growth, the amount of moisture in the root zone must be just right — too little moisture will reduce plant growth and development, restricting yield or causing the plant to die; if too wet, the soil is waterlogged, which could also restrict plant growth and development. The amount of moisture in the soil changes continuously, and the arrows in Figure I.1 show the main causes of this.

Evaporation

Solar energy vaporises water lying on the surface of the soil or plant foliage, releasing it to the atmosphere. Even in summer, this is only significant when the ground surface or canopy is wet, the top few centimetres of soil drying quite quickly.

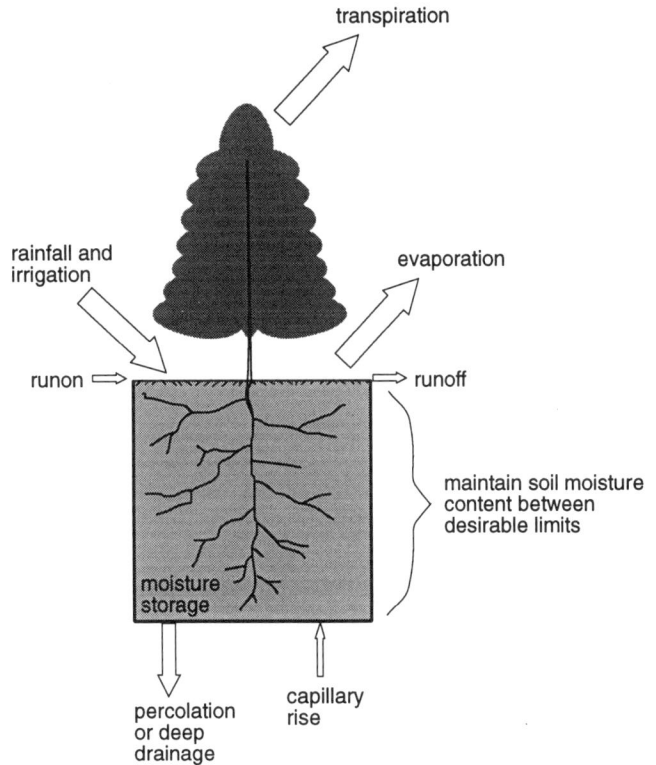

Figure I.1 *Water is added to and lost from the soil by a variety of mechanisms. Irrigation is designed to maintain soil water content between desirable limits*

Transpiration

This process is also driven by solar energy, which vaporises water held in the foliage of the plant, releasing it to the atmosphere. As water is lost from the foliage, it is replaced by moisture drawn from the soil by the roots. Consequently, transpiration continues even when the soil surface is dry, whenever the roots can still extract moisture from the soil.

This is the most significant mechanism that removes moisture from the soil and is discussed in much more detail in Chapter 3.

Runoff and runon

If water is applied to the ground surface more quickly than it can infiltrate, it will start to pond on the surface, and a stage will be reached where "runoff" will occur. Under natural rainfall, not much can be done other than to collect the runoff in drains and dams, and to control the rate of runoff to eliminate erosion. Under spray irrigation runoff of irrigation water is considered a waste of water, and therefore undesirable. In surface irrigation, water is deliberately flooded at the top of the bay or furrow and it runs toward the end of the block infiltrating as it goes, but water which runs off the end of the block should be collected for re-use.

Just as water runs off the surface to areas lower down, water can run on to the area from higher up.

Percolation

If the root zone is wetted, water will drain downwards under the action of gravity. If it percolates below the depth of the root zone, and the roots are already at their full extension, it becomes lost to the system. For irrigation water this would normally be considered undesirable, unless:

- The roots were able to keep growing and seek out the sub-soil moisture.
- Extra water was deliberately applied to wash accumulated salts out of the root zone (leaching).

Deep drainage will continue until moisture reaches an impermeable barrier, such as very heavy clays, or rock, where the rate of water movement through them is very slow, or a zone where the soil is saturated. The water surface at the top of this zone is the "watertable".

If these conditions exist near the ground surface (within 2 m), plant growth and development may be retarded, because root development will be restricted. Excessive irrigation, or less than ideal irrigation layout, can result in a significant amount of deep drainage. Over a long period of time, this can and has contributed to rising watertables under irrigation areas, resulting in significant problems in some locations. Some irrigation methods are worse than others in this regard, and good scheduling will also help minimise problems.

Capillary rise

This is only significant if the watertable is close to the root zone. Surface tension created in the film of water between soil particles can draw water upwards from the watertable, against the force of gravity. This would rarely contribute useful quantities under irrigation conditions, but can bring salt with it if the saturated zone is saline.

Rainfall and irrigation

These are the main processes which add water into the root zone. In some situations, there is no rainfall during the growing season, at least during dry years, so the crop is totally dependant on irrigation. This is a reasonably safe assumption when predicting the peak irrigation requirement during the planning stages of an irrigation project. In other situations, irrigation is used only to supplement the natural rainfall that can reasonably be expected during the growing season. This gives the ability to apply water at critical stages of growth, thereby maximising yield or quality of product.

Rain and irrigation are normally received at the ground surface (the exceptions are where the irrigation system is a subsurface one, or hydroponics). The water must infiltrate the surface, and move downward throughout the root zone. Consequently, a knowledge of water movement in soils is necessary. It is important to know the upper and lower limits to soil moisture content, and how to assess moisture content in the field. It is also important to determine how quickly water is being consumed by transpiration.

CHAPTER 1

Irrigation planning

This chapter looks at the many technical factors to consider before embarking on an irrigation project. Most are specific to the site; that is, each project needs to be evaluated for that particular situation, although industry guidelines are also available. The final decision will be greatly influenced by economic factors, outside the scope of this book, but good planning will assist accurate cost estimation.

For more details on dams, groundwater and reliability, refer to *Farm Water Supplies: Planning and Installation,* in this series.

Water supply

Sources

Irrigation water can be supplied from a variety of sources which can be classed as either regulated or unregulated watercourses, or a groundwater. In all but a few circumstances an irrigation license is required.

Regulated watercourses

Many Australian river systems have been dammed, and the water is released from the large storage reservoirs to provide a much more constant flow in the rivers. Water is either diverted to individual farms by various systems of weirs (to control river heights) and channels, or is pumped directly from the river (by individual farmers, or by a pump station supplying a group of farmers).

The conditions of the landowner's irrigation license provide for a fixed volume of water to be allocated each year. Some older arrangements allowed for any amount of water to be applied over a fixed area for irrigation. Town and industrial supplies, and irrigation of permanent plantations, are given priority over other irrigation uses during times of water shortage, and so have higher security of supply. During

Figure 1.1 *During high river flows, it may be permitted to draw water in excess of license allocations. These pumps can deliver water to a large earth tank, which provides an opportunity to expand the irrigated area in some years*

dry years, with low reservoir levels, allocations for low security users are reduced (in droughts, it may be as low as zero).

This usually results in irrigators reducing their irrigated crop area, although some farmers choose to maintain their normal programme in the hope that natural rainfall is sufficient to make up for reduced irrigation amounts.

Occasionally, during wet years when reservoir levels are high, water may be made available to irrigators in excess of their allocation. Some farmers build large water storage facilities, pumping into them when permitted, then being able to increase the area irrigated.

Regulation of river flows disturbs the natural flow regime, which has an impact on vegetation and wildlife in rivers and related wetlands, and can cause low flow salt concentration. Debate continues on the appropriate regulation policies to service environmental requirements.

Because it may take many days for water to reach a distant downstream property after it is released, irrigation schemes operate on a form of water ordering system. Usually this is easily managed, but sometimes problems can arise (for example, what to do if heavy rain occurs after water has been released but before it arrives at the farm) and sometimes on-demand irrigation systems (such as microirrigation) are difficult to provide for.

In all cases, the volume of water entering the farm is measured. A Dethridge wheel is used when the supply is from a channel, and an approved flow meter in the case of pipes.

Unregulated watercourses

Irrigation occurs from watercourses upstream of the major storage reservoirs, but the flow is unregulated and therefore

Figure 1.2 *Large earthen dam storing water for irrigation. Detailed planning is essential to ensure successful construction and reliable inflows. (Courtesy Central Highlands Management.)*

less reliable. Flows are usually at their lowest when irrigation demand is at its highest. On larger watercourses, direct pumping is an option. On small watercourses, greater reliability is achieved by constructing a storage on or adjacent to the watercourse. Water harvested during times of suitable flow is stored until required for irrigation, the size of the dam matched to the expected yield from the catchment of the watercourse and the water requirements of the crop.

An irrigation license is required from a watercourse, even though its flow is unregulated.

Bores and wells

Underground water may be available in sufficient quantity to provide irrigation supplies. Hopefully this would be available at a reasonably shallow depth, since the cost of drilling and operating a bore increases with depth. When attempting to locate ground water, utilise the hydrogeology services available through water supply agencies. A licence is also required to irrigate with ground water.

Bore pressure is sometimes sufficient to pump direct onto the field. If the yield is low, borewater would be pumped into a storage reservoir over a long time period, to provide a buffer for peak water requirements.

Dams

When planning dams, consider the following points:

- A detailed analysis of the amount and variability of the runoff generated from the catchment above the dam is required, so that the reliability of water supply can be assessed.
- A detailed site survey is necessary to estimate water storage and earthworks volumes.

Figure 1.3 *Bore water is directed into irrigation supply channel by this bore-head pump*

- Dams for storing irrigation water are usually large earthen structures. The suitability of the material at the site must be thoroughly assessed to ensure the stability of the dam, and to minimise leakage and seepage. Plans and specifications need to be matched to the mechanical strength of the material.
- Construction technique is critical to the success of the works.

Quantity

It is difficult to give general guidelines on the water requirements of irrigated crops, since they vary significantly between localities, and for specific site conditions. For example, heavy soils hold more water than light soils, so that the amount of water to be applied per irrigation, and the interval between irrigations, is different. Water consumption by a crop depends on its stage of growth; it is greatest when an actively growing plant canopy completely shades the ground surface, maximising interception of solar radiation. The loss of water by plants through transpiration depends on climatic factors: radiation, temperature, humidity and wind. Chapter 3 provides a detailed summary, with examples of calculating crop water requirements.

As an approximation, the water use of plants can be related to pan evaporation records, which are available as part of normal weather recording procedures at many localities. This approximation is acceptable for feasibility planning, where the time scale is measured over the months of the irrigation season, but is less accurate for time periods less than ten days or so. The average monthly evaporation, modified by the application of a "crop factor", less average monthly rainfall, is an indication of the crop water requirements for that period. A total for the months of the irrigation season is an indication of the seasonal requirement. Each site requires detailed analysis.

For initial planning, work on the basis of 2–8 ML/ha, the higher figure for summer crops in low rainfall areas.

It is necessary to consider the range of conditions likely to be experienced. In particular, the difference between average and extreme weather conditions, since this will partly determine the reliability of supply and the ability of the system to keep water up to the crop.

Quality

Poor water quality can cause a number of problems. Most of these are predictable, and standard tests are readily available.

IRRIGATION LICENCES

Irrigation licences are required for most irrigation projects in Australia. Although there are variations in procedures between states, the following summary generally applies. If approved, the licence will describe a number of conditions that must be complied with, in particular, the volume of water available and/or the area irrigated, depending on whether the watercourse is regulated or unregulated.

In addition, conditions may apply to the method of water extraction, the capacity or size of any equipment such as pumps, and the important dimensions of storage or diversion structures. It may also specify the streamflow limits when water harvesting must cease (such as a low water level at some measurable location).

A detailed proposal is required when applying for a licence. The proposal will be inspected for technical adequacy and possibly for environmental impact. It would normally be advertised, and public comment invited.

For some projects, specialised investigations may be required, because of the project size or the impact the proposal may have.

During 1995, a temporary embargo was placed on all new licence applications for the whole of the Murray-Darling basin, with some provision for exceptional circumstance. In 1996, this was extended, including some other catchments. It also restricts the transfer of entitlements, subject to a review of water policy. All storages are affected. It is now being reviewed.

Once this review is complete, it is likely that irrigation licences will be less readily available than in the past, increasing their value, with management and policy decisions made more at the catchment level, rather than statewide. "Sleeping" licences (held by landowners but not in use) may be re-allocated to areas of greater need. Water entitlements are likely to become tradeable between consumers.

The major concerns are summarised in Table 1.1, together with the various criteria which are measured, and the generally accepted range of values for each.

The electrical conductivity of the water (E C w) is a measure of the quantity of salts dissolved in it. Of particular concern is sodium chloride. Only small amounts of salt can be tolerated by most plants, although some are more salt tolerant than others. Table 1.2 provides further guidance to salt tolerance of various crops.

The amount of sodium can represent a toxic hazard to the plant, and can also affect soil structure. These effects are related not only to the amount of sodium present, but also to the amount of calcium and magnesium. The sodium absorption ratio (SAR) is a standard measure of the relative quantities of these components. A high percentage of chloride can also have toxic effects, particularly when sprayed over foliage, and particularly to vines and orchard trees.

Another aspect of water quality is the calcium carbonate saturation idex, which measures the relationship between pH, salinity and "hardness" (hard water is high in calcium and/or

Table 1.1 *Summary of water quality guidelines for irrigation. (Source: NSW Agriculture.)*

Type of problem Test	Unit of measurement	Degree of problem		
		None	Increasing	Severe
Salinity ECw	microS/cm	< 700	700 to 3000	> 3000
Poor internal drainage ECw Adj. SAR	microS/cm (mmole/L)$^{1/2}$	> 500 < 6	200 to 500 6 to 9	< 200 >9
Toxicity of specific ions to sensitive crops (a) Root absorption Sodium (as adj. SAR) Chloride (b) Foliar absorption Chloride	(mmole/L)$^{1/2}$ mg/L mg/L	< 3 < 140 < 70	3 to 9 140 to 350 70 to 150	> 9 > 350 > 150
Encrustation (scaling) CaCO$_3$ saturation index	–	< 0.5	0.5 to 1.5	>1.5
Corrosion CaCO$_3$ saturation index	–	> –0.5	–0.5 to –1.5	< –1.5
Acidity/Alkalinity pH	normal range: 6.0 to 8.5			

Table 1.2 *Salt tolerance of crops (in electrical conductivity of irrigation water, microsiemens per centimetre at 25°C). (Source: Awad, 1984.)*

Vegetable crops		Fruit crops		Field crops		Forage crops	
beets	2,700	olive	1,800	barley	5,300	tall wheat grass	5,000
broccoli	1,900	fig	1,800	cotton	5,100	couch grass	4,600
tomato	1,700	grapefruit	1,200	sugar beet	4,700	barley (hay)	4,000
cucumber	1,700	orange	1,100	wheat	4,000	perennial	
cantaloupe	1,500	lemon	1,100	safflower	3,500	ryegrass	3,700
broadbeans	1,500	walnut	1,100	soybean	3,300	trefoil, birdsfoot	3,300
spinach	1,300	peach	1,100	sorghum	3,300	phalaris	3,000
watermelon	1,300	apricot	1,100	peanut	2,100	tall fescue	2,600
cabbage	1,200	grape (cardinal)	1,000	rice (paddy)	2,000	crested	
potato	1,100	almond	1,000	corn (grain)	1,100	wheat grass	2,300
sweet corn	1,100	plum	1,000	sugarcane	1,100	vetch	2,000
sweet potato	1,100	blackberry	1,000	flax	1,100	sudan grass	1,900
pepper	1,000	avocado	700	cowpea	900	cocksfoot	1,500
lettuce	900	raspberry	700			trefoil, big	1,500
radish	800	strawberry	700			lucerne	1,300
onion	800					lovegrass	1,300
carrot	700					corn (forage)	1,200
beans	700					orchard grass	1,000
						meadow foxtail	1,000
						most clovers	1,000

magnesium). If the index is high, calcium and magnesium salts may form a crust inside pipes, which would be a major problem with some irrigation methods. If the index is too low, the risk of corrosion may be increased, since some crusting may also reduce corrosion.

Water high in iron may also cause problems. Iron can be dissolved in ground water, at a depth where there is little oxygen. In the presence of air, the iron oxidises, forming solid particles, discolouring the water, and leaving deposits on pipes. Deliberate aeration of water prior to its use, and settling the particles in still water, may be necessary and various aeration techniques are in use. A maximum iron concentration of 1 mg/L is recommended but problems can be experienced above 0.3 mg/L. Crusting of bore screens is one particular problem.

The presence of bacteria which use iron in their metabolism makes the problem worse, because they produce a rust coloured slime which can form on pipes, restricting water flows, and potentially blocking drip irrigation emitters, screens, and so on.

Water can sometimes contain small quantities of toxic trace elements and the accepted limits to these are summarised in Table 1.3.

Algae and bacteria can also be present, and may cause problems with microirrigation systems in particular. Chlorine treatment is necessary to keep pipes and equipment clean.

Conditions favourable to organism growth include high water temperature, and high concentrations of nutrients such as nitrogen and phosphorus, often associated with sediment carried in runoff flows.

There is increasing interest in using recycled wastewater for irrigation purposes, where algae, bacteria and high nitrogen levels may be present. For spray and surface irrigation of waste

Table 1.3 *Acceptable levels for trace elements. (Source: Awad, 1984.)*

Element	Level mg/L	Element	Level mg/L
aluminium	1.00	iron — spray irrigation	1.00
arsenic	0.10	iron — under tree irrigation	5.00
boron	0.30	lead	5.00
cadmium	0.01	manganese	2.00
chromium	0.10	nickel	0.20
cobalt	0.20	selenium	0.02
copper	0.20	zinc	2.00
fluoride	1.00		

water, aeration and exposure to ultraviolet radiation usually provide at least partial control over organisms in the water, but blockage risks within microirrigation will usually require chlorine treatment. There are restrictions on the choice of site for wastewater application (because of increased pollution risk), and the use of the product grown with it (because of possible health risks to consumers) but there is significant potential for re-use of this resource.

Farming practices can influence water quality in both dam water and watercourses. For example, nutrient, chemical and bacterial contamination can be caused by:

- runoff from paddocks,
- fouling by stock and wildlife,
- drainage from septic tanks,
- effluent outfall from sewage treatment works into streams.

As far as practicable, modify the source of contamination. Runoff from farming land is likely to have high nutrient loads, so long term attention to fertilising strategy is important. Fouling by stock can be reduced by fencing off dams and using troughs. Runoff from high risk sites (fertiliser dump pads, intensive livestock housing areas and livestock camps) can be diverted into holding ponds to avoid dams and streams. Vegetation filters and buffers can be installed. Divert clean stormwater runoff away from high risk sites. Where drainage water is collected, additional requirements should be considered (discussed in more detail later in this chapter).

Water testing

When planning an irrigation project, water quality should be assessed thoroughly, particularly for microirrigation.

Most regional centres have laboratories for assessing water quality. Tests can be arranged through agriculture or water supply authorities. Samples of around one litre are normally adequate, although some specialised tests may require a larger sample volume and/or a particular sampling technique. Specific information is best obtained from the laboratory. For reliable results when taking samples:

- Ensure that the container is absolutely clean.
- Rinse the bottle four times with sample water, disposing of the rinse water away from the sample site.
- Fill the container (leave minimum air space).
- Take the sample from a representative location (depth and location of the sample have a major impact on results).

- Label the container with all relevant details.
- Nominate the purpose the water is to be used for.
- Ensure the laboratory receives the sample as quickly as possible.

On-going quality testing may be necessary to check if changes occur over time. This may be necessary when pumping from bores or near estuaries, and where drainage or potentially contaminated water is collected or is diluted with clean water.

Soil

Soil conditions have an important influence over the success of irrigated production. Their physical, chemical and biological features will partly determine irrigation specifications, and will greatly influence irrigation management and ultimately crop performance.

Physical characteristics

The physical characteristics of soil largely determine the movement of water into and through the soil matrix (the infiltration rate at the soil surface, and the way water is redistributed throughout the root zone, under the action of gravity and soil suction forces), and the amount of soil water available to plants (the soil's water holding capacity). A full discussion of these characteristics is in Chapter 2.

Soils can be tested for their infiltration rate, and hydraulic conductivity of the soil at any location is a measure of its rate of internal drainage.

Soil texture is determined by analysing the proportions of sand, silt and clay particles. Soil structure is difficult to measure, but for a particular soil type, its bulk density may give some information on the volume of pores in the soil.

The behaviour of the soil when wetted should also be assessed. In particular, some soils have a tendency for slaking (a breakdown of the structure of soil crumbs when wetted) or dispersion (where the clay particles are released into suspension). These characteristics are undesirable, and contribute to problems such as surface crusting and poor internal drainage, which in turn result in poor crop establishment and performance, and ultimately difficult irrigation management. Laboratory tests can measure aggregate stability, percentage dispersion, and degree of crusting, but some simple field tests are possible. Place some dry soil crumbs into a glass dish containing clean water. Slaking will be observed if the crumbs appear to disintegrate and form a layer on the bottom of the dish. Dispersion will result in the water becoming cloudy.

These characteristics are partly affected by the chemical analysis of the water, particularly by the presence of sodium, and the SAR of the soil and of the water. Consequently, dispersion and slaking should be assessed both with distilled water and with irrigation water.

Management practices can influence these characteristics over time. For example correct irrigation management, appropriate cultivation and crop establishment methods, and the presence of soil conditioning agents such as organic matter will greatly influence soil characteristics. Gypsum is commonly used to overcome these problems, sometimes applied through irrigation water.

Some soils swell when wetted, and shrink and produce large cracks when dry. These will be difficult to manage, because initial water uptake will be high as the cracks fill with water, but this will drop substantially and quickly when the soil swells and the cracks close up. The cracks in these soils are readily observable, but laboratory tests are used to measure the swelling percentage.

Soils may have additional physical features which impede irrigation development or management, such as rocks, stumps, and subsoil barriers.

Chemical and biological characteristics

Testing of soil fertility (nutrients, pH, SAR and so on) is a standard management practice for all farms, and sampling methods, and test procedures and interpretations, are available from government and industry agronomists. Fertiliser practices are outside the scope of this book. The following comments summarise some of the factors to consider with irrigated enterprises.

Soil testing for nutrient status may not reveal all that needs to be known about the soil. In particular, many of its physical features are not measured with standard tests, or require more specialised tests at extra costs. Sampling for soil fertility may involve the mixing of many small subsamples from throughout the field, which will mask variability within the field. Most samples are taken from the topsoil, but subsoil conditions need examination as well.

Biological activity in soils is of considerable importance, yet often overlooked when assessing soil characteristics. A wide range of macro- and micro-organisms can influence soil conditions such as structure and nutrient availability. The density of their populations and activity levels are dependant on a wide range of soil conditions, but there is a strong relationship to organic matter and pH. Management practices have a large influence over these factors.

Variability of soil characteristics

Soil characteristics may vary between different locations within a field, and vary with depth. Soil characteristics should therefore be assessed in sufficient locations, at adequate depth at each location, to ensure that this variability has been measured.

Variability at the surface can be detected by taking a large enough number of samples. Variations can sometimes be linked to changes in topography or vegetation, but more commonly by soil colour. For more intensive projects such as horticultural crops, sampling could be conducted on a grid throughout the field, which will facilitate mapping of soil type variations.

At each sampling site, an inspection or sample of the soil to at least the full depth of the root zone should be conducted. This can be done by core sampling or by digging pits with a backhoe. A shovel or auger can be used, but these will mix soil from different depths, which may give less useful information. Such samples will identify soil characteristics throughout the root zone, and the presence of subsoil conditions such as hard pans, underlying rock, and heavy subsoils with low hydraulic conductivity. Observe the depth of root penetration of existing vegetation, and the depth at which changes to soil type or colour occur.

Because soil characteristics influence irrigation management, irrigation blocks should be matched to changes in soil characteristics, particularly soil water holding capacity, but also topography in the case of surface irrigation methods. In this way, irrigation can target crop water requirements more accurately, and irrigation is likely to be more uniform and with higher application efficiency.

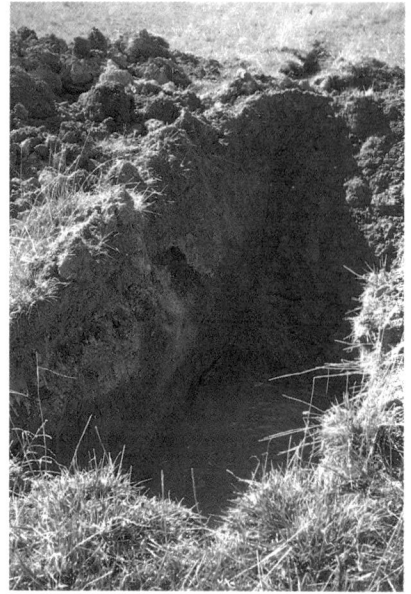

Figure 1.4 *A series of test pits such as these can be used to inspect soil conditions throughout the depth of the profile. In this pit, note the light coloured silty material just below the topsoil, and heavy clay material in the subsoil. The water in the base of the pit may be caused by a high water table or by poor internal drainage*

Topography

Slope and the changes in slope of the irrigated area have a major influence over selection of the method of irrigation, and the specifications of the method selected. For example, slope along a bay or furrow will partly determine the preferred length of run for surface irrigation methods, because of its influence over the travel speed of irrigation water over the ground surface. Likewise, it has a similar influence over the capacity of supply channels and drains.

For spray and microirrigation methods, the slope along pipelines will partly determine the water pressure, and therefore output. Differences in elevation influence the power required by pumps to lift water.

There is also a link between site topography and the management of stormwater runoff. The size of the catchment areas generating runoff will determine the volume of water to be managed, and topography determines the direction of drainage patterns for the area, whether these are within or outside the irrigated area.

A detailed topographic survey is essential for detailed planning of an irrigation project. The survey needs to include water storage sites, pump station locations, natural features such as waterways and timbered areas, and farm infrastructure such as roads, power lines and dwellings. It needs to include stormwater catchment areas, (perhaps to a separate scale) so that runoff volumes can be predicted, and decisions made on whether the runoff is harvested into dams or diverted around the irrigated area.

Stormwater management and tailwater recirculation

This needs to be considered from a number of points of view. In some cases runoff is harvested and stored in dams, and represents the source of water for irrigation. Detailed catchment and climatic studies are required to determine the expected yield of the catchment and its reliability, which will determine the required storage size and the area of irrigation it can support. For surface irrigation layouts, the stormwater runoff generated from the irrigated area, if collected, can represent a significant contribution to the total crop water requirement.

In other cases the runoff generated from the irrigated area could be considered a pollutant under environment protection guidelines, not to be discharged into natural watercourses, because of the risk it will contain chemicals and nutrients. This water should be stored on-farm, and hopefully re-used as irrigation water. This may not be possible, either because there is no infrastructure to provide for storage and recirculation, or because the volumes of runoff are too large to store. Under these circumstances, the system should aim to contain at least the initial part of runoff, and certainly that from high risk sites (such as intensively managed areas) where the concentration of contaminants is likely to be greatest. Management practices such as vegetated strips and artificial

Figure 1.5 *Tailwater from this border check irrigation system is delivered to a sump. From here, tailwater is pumped to a spray irrigation system, minimising environmental problems, and maximising use of water. (Water can also be supplied direct to the spray irrigation pump.)*

wetlands can be used to partially treat stormwater flows and buffer storage can be utilised, and runoff can be diverted around high risk areas to minimise problems.

Surface irrigation layouts should aim to collect and re-use all tailwater (irrigation water draining from the end of bays or furrows). It should not enter the natural drainage system, for example, through roadside drains, or through connected river system features such as billabongs. Tailwater should be collected in a system of drainage channels. The layout could be designed so that it can be easily pumped back into the irrigation storage for re-use. A well-designed layout will minimise the amount of tailwater.

The issue of soil erosion should be considered. The system should be designed so there is minimum (zero) soil loss, whether from irrigation practice or stormwater flows. Bare soil should be protected by vegetation or mulch, tillage practices should minimise disturbance, and water flow should occur at non-erosive velocities.

System design capacity

The variability of the weather patterns that determine crop water requirements make predictions very difficult. It is important to distinguish between average water requirement and peak water requirement. Within a season there will be a period when water requirements are at their maximum. This is when the plant canopy is at its maximum, and evaporation demand is at its greatest. Although this might only last for a few days or weeks the irrigation system must be able to cope with this situation. Further, there will be some seasons which impose a greater water requirement than others. The worst case occurs during major droughts. It needs to be decided whether the irrigation system must be able to cope with the worst case scenario. If so, the supply of water must be extremely reliable, and the capacity of the equipment used to reticulate the water must be sufficient. During average or wet years, and for most of dry years, the system will not be working to its capacity. An alternative is to design for a lesser capacity, and tolerate the yield loss that is likely during extreme dry periods. The final decision will depend on:

- The capital and operating costs of the alternatives.
- The likelihood, amount and value of yield reduction.
- The reliability of the water supply.

Irrigation infrastructure (pumps, pipelines, channels) is therefore designed and selected with a maximum capacity to meet the pre-determined peak irrigation requirement. This limits its ability to keep up with conditions that might be worse than expected, and emphasises the point that careful planning of peak requirements is essential. In addition, it is useful to determine the range of operating conditions (not just the peak), since some equipment (particularly pumps) operate efficiently over a relatively narrow range of duties.

There are circumstances that will enable an irrigation system to cope better under extreme conditions:

- Soils which have a high proportion of available water, and crops which have a well developed root system, can have a longer interval between waterings, enabling equipment of fixed capacity to complete each irrigation cycle satisfactorily.
- Final crop yield is often most affected by soil water status at particular stages of growth. If supply or equipment capacity is limited, irrigations should target these.
- Irrigators should have a contingency plan for equipment breakdown during critical stages. Correct maintenance will keep equipment in good working order, and ensure pipes and channels are able to work to their capacity.
- Irrigators may need to concentrate a limited quantity of water onto a reduced area of crop in extremely dry periods.

Application efficiency

Water application efficiency refers to the amount of irrigation water that becomes available to the plants' root zone, as a

Figure 1.6 *Strong wind is blowing a large proportion of water off this spray-irrigated field, the wetted area reaches only one bed width to windward*

percentage of the amount of water pumped on farm or delivered to the farm boundary. In this context, application efficiency is reduced by the loss of water from:

- Wind blowing spray droplets away from the site or pushing water to one side of a bay.
- Evaporation, of droplets in the air and of free water lying on plant and soil surfaces.
- Runoff, when application rates exceed the soil infiltration rate.
- Deep percolation, where irrigation water infiltrates below the root zone, and there is no later possibility of root growth to extract it.

Most of these factors are related to the design of the system or its management.

Because the cost of water depends on the volume received or pumped, and not on the volume used by crops, a poor

Figure 1.7 *Runoff of irrigation water can be seen flowing from this orchard*

application efficiency will contribute to a lower economic return, as costs will be higher than they need to be. On the other hand, to correct or provide for high application efficiency may require higher initial expenditure (landforming, construction of tailwater recirculation systems and so on), but this is not always the case. Attention to detail regarding soil types, and appropriate scheduling techniques will improve application efficiency.

Poor application efficiency is a contributor to overwatering, which can exacerbate water logging and salinity problems. (Note that application efficiency is not the same as water use efficiency, which is normally defined as the ability of the plant to use water in its growth and development. Another measure of efficiency is mechanical efficiency, the amount of water applied per kilowatt of energy used at the pump.) Application efficiency for microirrigated systems should be as high as 90–95 per cent. For spray systems, application efficiency is typically 70–80 per cent, but lower in windy conditions.

Under surface irrigation application efficiency is typically 50–70 per cent, but efficiencies as low as 30 per cent have been recorded.

When calculating the volume of water to be applied during an irrigation, the crop water requirement is divided by the application efficiency, giving the amount to be delivered.

Uniformity of application

It is desirable that all plants within the field receive the same amount of water, and this is the objective to start with at the design stage. This is not always easy to achieve, because of the nature of surface irrigation methods, the circular wetting patterns of sprinklers, and pressure and head variations along pipelines and channels.

Uniformity of application can be enhanced by correct design procedures, nominating acceptable pressure variation and matching water application to soil condition. Note that uniformity of application is observed and measured at the soil surface, and it is not always clear how lateral infiltration may affect uniformity with depth through the root zone.

Correctly designed microirrigation systems have a high uniformity of application, a significant advantage of such systems provided the system is designed with an acceptable pressure variation (about 10–20 per cent). Spray irrigation systems can have high uniformity as well, but it is influenced by the wetting pattern of individual sprinklers (which is partly determined by sprinkler spacing). Surface irrigation methods typically have a low uniformity of application, but this can be improved with design details and management strategies.

Watertables and salinity

An increasingly large amount of irrigated land is affected by high (less than 2m from the ground surface) or rising watertables. In some areas, production is severely limited because the watertable restricts root development, even rising to the ground surface. A compounding problem is that much of this water is saline, further hampering plant production. In some areas, drainage and seepage from these sites affects adjoining land, and enters the natural drainage system affecting watercourses.

For areas characterised by heavy subsoils, it is necessary to distinguish between the true groundwater table, and one which may be perched on the subsoil layer, perhaps temporarily. Perched water is usually of reasonable quality.

A number of strategies could be considered to manage watertable levels, and therefore partly manage salinity hazards. A number of on-farm drainage methods are possible, which are described in Chapter 9. These have the potential to remove groundwater and thereby lower the watertable, but the drainage water is likely to be saline and therefore not allowed to be discharged off-farm. If the water is of reasonable quality, it could be used as irrigation water, or perhaps blended with high quality water to enable its re-use. If it is too saline for re-use, it may be necessary to deliver the water to an evaporation basin, which is often done on a regional scale.

It is better to reduce the amount of water percolating to the watertable in the first instance. Irrigation of light, sandy soils presents the greatest risk. For surface irrigation methods, correct bay or furrow slope, matched to the length of run and soil intake characteristics will greatly assist, by providing a more uniform and efficient application. Irregularities in the slope within bays should be avoided, to reduce the risk of water ponding after irrigation ceases. Laser operated earthmoving machinery is employed to provide the necessary slopes during initial construction, followed by bay maintenance as required. Older style layouts can be reconstructed with better and more uniform grades, whilst also providing more cost effective layouts for machinery and labour. Seepage from channels should be minimised.

Spray and microirrigation methods are better able to match water application to crop water requirements, and therefore reduce water movement to the watertable, provided application rates are chosen carefully. For all systems, accurate and frequent monitoring of soil water status (see Chapter 2) will enable irrigators to reduce overwatering. Trees and other perennial vegetation have the potential to consume large quantities of water, and may assist in watertable control.

In some situations, it may be necessary to leach accumulated salt from the root zone, by applying more water than the crop requires. The leaching requirement is expressed as a percentage of the irrigation requirement, depending on the amount of salt to be removed, the salt content of the irrigation water, and the amount of rainfall that could be experienced during the irrigation season. It will only be effective if the watertable is at a sufficient depth to receive the extra water without affecting plant growth.

Irrigators are encouraged to monitor the depth to the watertable on their farm using testwells or piezometers. A simple test well, made from PVC pipe, will show watertable depths, and can facilitate water sampling for salinity testing.

Selection of irrigation method

A wide range of irrigation methods are available, classified as either surface (sometimes called flood or gravity), spray or microirrigation methods. They are described in detail in Chapters 5, 6 and 7, which summarise the key features and common applications from a technical point of view.

In addition, factors such as soil characteristics and water supply, described earlier in this chapter, need to be considered.

For some projects, the choice of irrigation method is obvious, because the technical constraints dictate. However, for some projects the choice is not so obvious, and a wider range of factors will influence the decision:

- Funds available to invest in irrigation hardware.
- Operating costs of labour, fuel and maintenance.
- Ability to apply smaller quantities of water at more frequent intervals.
- Availability of labour, and the possibility of automation (even in part).
- Skills of the labour force and management.
- The area irrigated and the scale of operations.
- The value of the crop, its sensitivity to irrigation management at critical growth stages, and the probability of rainfall during the irrigation season.
- Personal preference and experience.
- Additional purposes, such as fertiliser application or frost control.

Project management

A substantial amount of planning is needed to initiate an irrigation project. The need for an accurate and detailed topographic survey, and thorough assessment of crop water requirements and soil characteristics have already been mentioned. An irrigation concept plan can be established by integrating these requirements with those of the preferred irrigation method and its management, over the whole farm. A base map or aerial photograph can be overlaid with a soils map and irrigation layout. Detailed design and project costing can then follow, fine tuned as required.

The design should be completed by a competent, preferably certified, irrigation designer. For example, it is necessary to determine the optimum pipe diameter and pump sizes based on both capital costs and annual operating costs, and to consider additional criteria such as water velocity. These need to be assessed under a range of conditions.

Evaluation will be needed for development, licence, environmental and budget approvals. The degree of detail required will partly depend on the scale of the project, its likely impact and the requirements of the approving agencies.

Tenders or quotes for installation can be called for if the project is approved. The more formal tendering procedures require a fully detailed set of plans and specifications to be available, as well as a copy of the contract. Although appearing to contain a lot of paperwork, these procedures are recommended to ensure competitive and equitable tendering, to the benefit of all parties. Calling for quotations may provide more opportunity for negotiation with contractors, but at the risk of less control.

The contract between owner and installer should refer to the agreed plans and specifications, with variations accepted only by agreement between the parties. It should also refer to minimum standards of product manufacture, installation and testing, available from various industry and government authorities, for equipment and earthworks. The contract should refer to procedures to follow in the event of a dispute, the schedule of inspections and progress payments, and may include timeliness clauses.

Provide ample time for the installation of the system, allowing for wet weather delays, ordering and delivery of materials, and thorough testing of the system, so that the system can be run effectively when required at the commencement of the season.

CHAPTER 2

Soil water

S oil conditions strongly influence the amount of water available to plant roots, and the movement of water through the soil. A thorough assessment of soil conditions is an essential early step in irrigation planning, and successful irrigation management.

Soil texture

This term refers to the relative proportion of sand, silt and clay in the soil. Each of these components is defined by its size (sand 0.02–2.0 millimetres, silt 0.02–0.002 millimetres, clay less than 0.002 millimetres). The soil texture is more correctly referred to as the soil's particle size distribution.

This is not the same as soil structure, which describes the way the soil particles are aggregated together (in crumbs or blocks). Structure also affects the way water behaves in soil, but this is discussed later.

The main soil types can be classified according to their texture. Figure 2.1 shows a commonly accepted classification using the textural triangle derived in the USA.

To determine soil texture, a sample of soil is oven dried, weighed, crushed and passed through a set of sieves of particular mesh sizes. The amount of material retained by each sieve is weighed. Silt and clay (the fines) both pass the smallest sieve and are collected in a tray, and are separated from each other in a different way. The sand and coarse fractions are weighed and recorded as a percentage of the initial weight.

The proportion of silt and clay are measured during a hydrometer test. A hydrometer is a simple floating apparatus that measures liquid density. If the silt and clay are thoroughly mixed with water in a measuring cylinder, and then allowed to stand, the silt particles, being heavier, will gradually sink to the bottom, whilst the clay particles will remain in

33

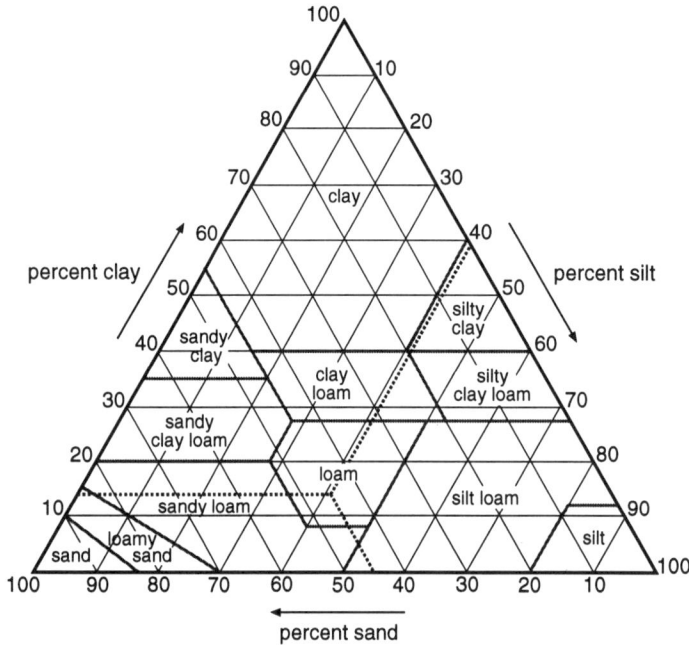

Figure 2.1 *Particle size distribution determines soil texture (Doneen, L.D., 1971)*

suspension. After a certain time of settlement, the hydrometer is used to measure the density of the liquid, which is a measure of the amount of clay in suspension, and therefore of the clay content of the soil sample. What is left must be the percentage of silt.

In the field, an experienced person can assess soil texture according to how a sample of soil "feels" in the hand. Some guidelines are contained in Table 2.1.

Soil texture has an important part to play in the behaviour of soil water. The arrangement of soil particles also determines the arrangement of the spaces between soil particles; the pore space. The pore space is where soil water and air are contained. Large pore spaces drain more readily than small pores. Small pores hold water more tightly because the surface tension of the water film in the pore is more able to resist the force of gravity trying to move the water downward through the profile.

Heavy soils, those containing a large proportion of clay particles, have a large proportion of small sized pores, so hold a relatively large amount of water against gravitational drainage, compared to lighter soils containing a larger proportion of sand.

Consequently, the physical make-up of the soil matrix plays an important part in soil water relations.

Table 2.1 *Field guide to soil texture*

Sand
Sand is loose and single grained, and the individual grains can be seen or felt. Squeezed in the hand when it is dry, it will fall apart when pressure is released. Squeezed when moist, it will form a cast, but will crumble easily when touched.

Fine sand
Fine sand is a loose soil in which a large proportion of individual grains can be seen or felt. Squeezed in the hand when dry it has a slight tendency to cohesion. Squeezed when moist, the cast will bear only very careful handling, otherwise it will crumble.

Sandy loam
A sandy loam is a soil containing much sand but which has enough silt and clay to make it somewhat cohesive. The individual sand grains can readily be seen and felt. Squeezed in the hand when dry, it will form a cast which will readily fall apart; but if squeezed when moist, a cast can be formed that will bear careful handling without breaking.

Fine sandy loam
A fine sandy loam contains a fairly large proportion of finer sand which gives the soil a gritty feel. Silt and clay make the soil rather more mellow to the feel than a sandy loam. Squeezed in the hand when dry it will form a cast which needs very careful handling. Squeezed when moist the cast will break fairly readily.

Loam
A loam is a soil having a mixture of the different grades of sand, silt and clay in such proportion that the characteristics of none predominates. It is mellow with a somewhat gritty feel, and when moist is slightly plastic. Squeezed in the hand when dry, it will form a cast that will bear careful handling, while the cast formed by squeezing the moist soil can be handled quite freely without breaking.

Silt loam
A silt loam is a soil having a moderate amount of the fine grades of sand and only a small amount of clay, over half of the particles being of the size called "silt". When dry it may appear quite cloddy, but the lumps can be readily broken; and when pulverized it feels smooth, soft and floury. When wet the soil readily runs together. Either dry or moist, it will form casts that can be freely handled without breaking. When moistened and squeezed between thumb and finger it will not "ribbon", but will give a broken appearance.

Light clay loam
A light clay loam forms clods or lumps which are fairly hard when dry and the clods can be quite readily broken. When the soil is moist, it will barely form a "ribbon", but it will break extremely easily.

Clay loam
A clay loam is a fine textured soil which usually breaks into clods or lumps that are hard when dry. When the moist soil is pinched between thumb and finger, it will form a thin "ribbon" which will break readily, barely sustaining its own weight. The moist soil is plastic and will form a cast that will bear much handling. When kneaded in the hand, it does not crumble readily but tends to work in a heavy compact mass.

Heavy clay loam
A heavy clay loam will form hard lumps or clods when dry and when wet the soil will form a "ribbon" which will bear some handling and will tend to cling together. It could be distinguished from a clay in that the "ribbon" will tend to break up.

Clay
A clay is a fine textured soil that usually forms very hard lumps or clods when dry and is quite plastic and usually sticky when wet. When the moist soil is pinched out between the thumb and finger, it will form a long flexible "ribbon". Some clays very high in colloids are friable and may lack plasticity at all conditions of moisture.

Source; New Zealand Standard 5103:1973

Soil structure

Soil structure refers to the way soil particles are aggregated together to form crumbs, blocks, or clods. Soil structure depends partly on the soil texture, but also on other factors such as the amount of organic matter in the soil, the population and activity of soil organisms (for example, earthworms), cultivation practices, even the way rain or spray irrigation droplets impact on the soil surface. These factors are related to management, and are variable over time. Soil structure can therefore be short term, temporary property of the soil, whereas soil texture is permanent.

Soil structure will influence water intake rates at the surface, which is to be encouraged in heavier soil types, or those which tend to develop a surface crust. Incorporation of organic matter is a useful practice for this purpose (as well as having other beneficial effects). The application of mulch to the ground, should also be considered. Practices such as excessive cultivation, cultivation when too wet or too dry, application of excessive amounts of irrigation water, and heavy spray irrigation on sensitive bare soil, can cause a deterioration of soil structure.

Soil structure will also influence water movement down the profile after infiltration. A well-structured soil will drain relatively quickly after heavy rain or irrigation, and so minimise the problem of temporary waterlogging. Although draining freely, a heavy textured but well structured soil will still hold adequate quantities of water available to plant roots. Poorly structured soil may inhibit plant growth and development by being waterlogged in wet conditions and by restricting the development of the plant root system.

When assessing soil conditions, do so through the full depth of the profile, to at least the depth that the root system of the proposed crop is expected to penetrate. Good conditions at the soil surface can hide subsoil problems. Soil characteristics can vary markedly over short distances, so assessments need to be conducted over a number of sites throughout the field.

Water in the soil matrix

Figure 2.2 attempts to illustrate soil particles in the soil matrix, and the gases (air and water vapour) and liquids (water, with dissolved material) that occupy the pore spaces between particles. It is not a particularly accurate drawing, because of the many and varied sizes and shapes of soil particles, but sufficient to describe the way water is held in the soil.

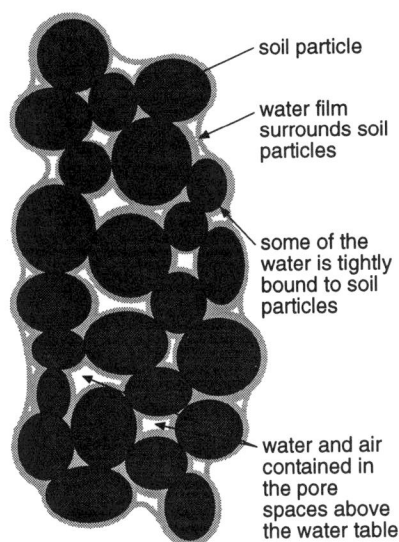

soil particle

water film surrounds soil particles

some of the water is tightly bound to soil particles

water and air contained in the pore spaces above the water table

Figure 2.2 *Arrangement and size of the pore spaces in the soil determine soil water holding capacity. Water is held in the pore space by surface tension and other forces*

A film of water surrounds the particles and occupies a portion of the pore space, firstly because of attraction between water molecules and the soil particle (hygroscopic water), but also because the surface tension of the water film holds it in the pore space against the force of gravity (capillary water). The remaining pore space can also hold water, after heavy rainfall or irrigation, but this will drain downwards under gravity, unless prevented by an impermeable barrier or the watertable (gravitational water). If not containing water, this remaining pore space will contain air and water vapour.

The relative proportion of these components of soil water varies with soil texture, since this will influence the size and nature of the pore space. Consequently, soil texture plays a critical role in determining the amount of water available to plants.

Gravitational water, although present in the root zone temporarily, is considered unavailable to plants, because it drains away in a relatively short period of time. In any case, excessive amounts of moisture will reduce the amount of air in the root zone, and conditions will be waterlogged. The term "field capacity" is commonly used to describe the soil water content of a wetted soil after 24 to 48 hours drainage. This point is not accurately defined, but occurs when rapid drainage of gravitational water ceases, after which much slower water movement occurs.

Hygroscopic water is bound quite tightly to the soil particles, so that plant roots cannot extract it, and it is also considered unavailable. Consequently, capillary water is that part of the soil water content that is available to plants. It is held in the soil matrix enabling plant roots to absorb it. However, as water content falls, in response to evaporation and transpiration, the thickness of the water film in the pore space diminishes, and surface tension forces become more dominant. Plant roots must work harder to extract the remaining water, against the attraction that the soil exerts on the water. A stage is reached during drying when plant roots cannot exert enough effort to remove water against soil suction, and the plant wilts. If prolonged, the plant will not recover, and will die. The moisture content at which this occurs is referred to as "permanent wilting point".

Field capacity and permanent wilting point therefore define the limits of soil water available to plants, and the difference between them is known as "available water". In irrigation, it is necessary to apply water well before wilting point; as plants struggle for water, their growth and development is affected even if they survive the dry spell. An important management decision is deciding when to irrigate, and this is discussed further in Chapter 4.

The concept of soil suction is a quite well defined term, and can be measured quite accurately. It is equivalent to the amount of vacuum required to suck water out of the soil against surface tension forces. Permanent wilting point is generally accepted to be the moisture content at a soil suction of 15 bars (one bar is equal to atmospheric pressure), although this does vary between different plant species and soil types. Field capacity is less readily defined, but may be at a suction between 0.1 and 0.3 bars.

Measurement of soil water content or soil suction

The amount of water in soil can be expressed in a number of ways:

- As a percentage by weight; grams of water per 100 g of soil.
- By the depth of water per metre of soil (that is the depth of water that would occur if all the soil water was extracted and spread over the surface of the soil) (Figure 2.3). This is a particularly convenient method of describing soil water content, because it is expressed in the same way that irrigation (and evaporation) amounts are normally measured.
- As the pressure of soil suction: that is, in bars.

Testing methods

Sampling and gravimetric measurement
A sample of soil is weighed, dried in an oven at 105°C (not so hot as to burn organic matter) for 24 hours, then re-weighed. The weight loss is the amount of water evaporated.

This is the traditional method of measuring water content, but is too slow and inconvenient for field use. Many samples would be required to give useful field information, at different depths in the root zone, and drying facilities are required.

Sampling using experience
Many irrigators can tell the moisture content of their soil by how it feels in the hand, when rolled into a ball or ribbon. Table 2.2 is a summary, but it is by no means an accurate assessment. Sampling tends to be done in the topsoil, which is not necessarily representative of soil water conditions in the subsoil.

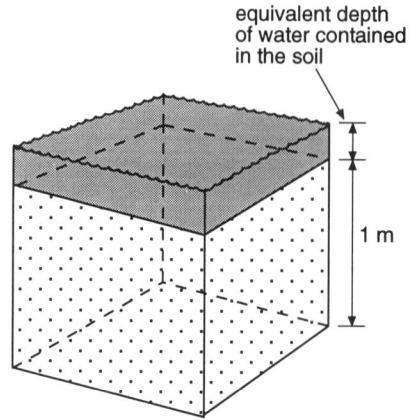

Figure 2.3 *Soil water content is often expressed as mm of water per m depth of soil, as if all the soil water was extracted from a 1 m cube of soil and spread over the top*

Table 2.2 *Field guide to soil water content*

Per cent available moisture remaining	Loamy sands and sandy loams (coarse textured)	Very fine sandy loam and silt loam (medium textured)	Silty clay loams and clay loams (fine textured)
0–25	Dry, loose, flows through fingers	Powdery, sometimes slightly crusted, but easily broken down into a powdery condition	Hard, baked, cracked; difficult to break down into powdery condition
25=50	Appears to be dry, will not form a ball with pressure	Somewhat crumbly, but holds together with pressure	Somewhat pliable, balls under pressure
50–75	Tends to ball under pressure, but seldom holds together when bounced in hand	Forms a ball, somewhat plastic, sticks slightly with pressure	Forms a ball, ribbons out between thumb and forefinger, has a slick feeling
75–100	Forms a weak ball, breaks easily when bounced in hand, will not slick	Forms a very pliable ball, slicks readily	Easily ribbons out between thumb and forefinger, has slick feeling
100 (Field Capacity)	Upon squeezing no free water appears on soil, but wet outline of ball is left on hand; soil sticks to thumb when rolled between thumb and forefinger	Same as sandy loam	Same as sandy loam
Saturated	Free water appears on soil when squeezed	Same as sandy loam	Same as sandy loam

Source: New Zealand Standard 5103:1973

Gypsum blocks

Two electrodes with wires attached are embedded in a small block of gypsum, which is then buried in the soil so that soil water penetrates the block. The electrical conductivity between the electrodes is measured, which will vary with the soil water content. Unfortunately, it also varies with solutes dissolved in the soil water. The readings obtained must be calibrated for each soil type, so initial gravimetric measurements are required. Care must be taken that the location and method of installation of the block truly represents conditions in the root zone. Multiple blocks are easily installed at different locations and depths in the field read by a portable meter. There are alternatives to gypsum .

Neutron probe

This is a much more sophisticated, and expensive, apparatus, but one which gives much more useful information. The apparatus consists of a radioactive source of fast neutrons, and a neutron counter, which are lowered down an aluminium access tube into the root zone. When fast neutrons strike hydrogen atoms (found in the soil and soil water, H_2O), the impact can absorb their energy. The number of neutrons scattered back toward the source are counted, which will be related to the soil water content.

This system has a number of features:

- A single apparatus is used for many access tubes. (Most farmers employ a consultant to bring the apparatus, take the readings, and interpret the information.) There is little problem in taking readings at many locations in a field. Only the top of the access tube protrudes at ground level, so it is easy to protect.

- Multiple readings are taken at different depths at each location, giving soil water information throughout the depth of the root zone, and beyond if necessary. This is particularly useful, as it enables estimates of water remaining in the profile.

- Readings are taken frequently throughout the season, usually at five or seven day intervals. Readings show water content at the time, so water content can be monitored. Knowing this at each depth through the profile gives an indication of where roots are active, and how water moves through the soil. Sequential readings enable the timing of the next irrigation to be determined.

- Data is logged with the apparatus, requiring down-loading to a computer. Special software is used to interpret the readings, and give a print-out of water content versus depth. Readings must be calibrated for the site. No cabling is required between stations.

- The neutrons are emitted from the source in all directions, so readings apply to a spherical zone of soil surrounding the source.

- The access tube must be installed without disturbing the soil matrix surrounding it.

- Being a radioactive source, certain precautions are required in using and handling the equipment, and the operator must be licenced.

Figure 2.4 *Typical neutron probe application.* Top left, *field monitoring cotton;* top right, *monitoring in a vineyard under drip irrigation;* bottom left, *augering for access tube installation;* bottom right, *data transfer to computer. (Courtesy Irricrop Technologies P/L.)*

EnvirosSCAN® capacitance probe

This apparatus measures the di-electric constant of the soil, which is proportional to the soil water content. A number of special sensors are mounted on a rigid probe, their position corresponding to various depths in the root zone. The probe is then permanently located in an access tube. Multiple tubes are required in each field to give representative results. Each tube site is connected by cable to a central data logger, for later analysis by computer.

The advantage of this system is that readings can be taken at any time, so soil water content can be monitored continuously. This can be of importance in high value drip-irrigated crops when the irrigation system can be accurately controlled, and irrigation management can respond to crop needs by the minute if required.

Figure 2.5 *EnviroSCAN® instrumentation removed from its access tube in potatoes*

Heat pulse probe

A heat source and temperature sensor are mounted into a small steel probe. A pulse of heat is emitted, and the rate of temperature change is measured. This is related to the soil water content.

Time delay refractometer

This measures the electrical response to a fast electrical pulse sent to special probes along a wire, which is also affected by the di-electric constant of the soil and therefore its water content.

Tensiometer

This instrument measures soil suction rather than water content (or a factor related to water content). It consists of a clear tube with a porous ceramic tip. The length of tube is selected to enable the tip to be located at the preferred depths in the soil (usually 30, 60 and 90cm at a number of sites within the field). The top of the tube contains a filling cap and a suction gauge. The tube is initially filled with water. As soil surrounding the tip dries, it will absorb water from the tip, which will create a partial vacuum in the tube in proportion to soil suction, and this is measured directly by the gauge.

As with any instrument, its location must be chosen to represent soil conditions in the root zone, and not be located, say, under a dripper where readings would be quite misleading. Hence the need for multiple tensiometers. Three depths are recommended at each site, so that moisture changes throughout the root zone can be monitored. Installation technique is also important. The top of the instrument protrudes above ground level, so it needs to be protected against damage.

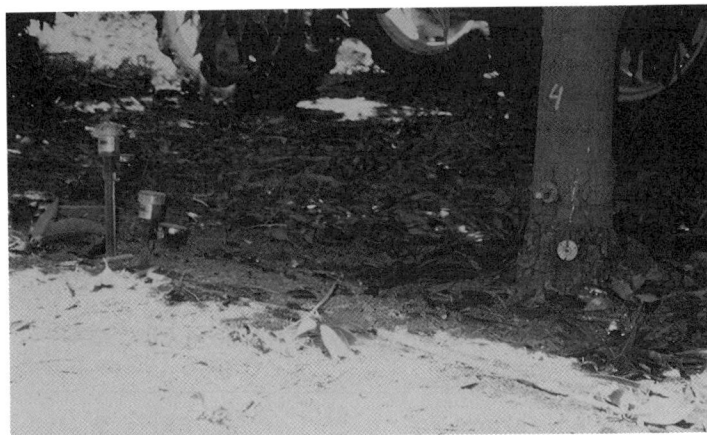

Figure 2.6 *Tensiometers installed under mangoes*

The root zone

The amount of soil water available to a plant depends on the depth of its root system, and the density of active roots throughout that depth.

Plants which have a deep root system have access to a larger volume of soil, and therefore a larger volume of soil water, than do shallow-rooted plants. Table 2.3 is a guide to the expected root depths for some selected crops. More specific advice should be sought from regional agronomists.

The data in Table 2.3 shows a wide range of values for each crop. This is the case in the field, since a large number of factors will influence root depth and distribution. In very heavy soils, root depth will be shallower than in light soils, partly because of the difficulty roots may experience in penetrating harder soils, but also because heavier soils hold more water than light soils, so roots may not need to extend as far to meet their water requirements. Any barriers to water movement or root penetration, such as a hard pan or high watertable, will obviously reduce root depth.

For annual crops, Table 2.3 refers to the root depth once the crop is fully matured. The root system will take some time after initial establishment to reach its full depth, which has important implications to irrigation management discussed later.

Table 2.3 also assumes the crop is adequately supplied with water. Plant roots will advance considerably to seek moisture as a soil dries, but under irrigation, it has less need to do this. However, it is still desirable to establish a substantial root volume under irrigation, to maximise the plant's access to soil water (and nutrients), reducing dependence on frequent waterings, and ensuring adequate anchorage of the plant.

Table 2.3 *Expected root depths under irrigation*

Horticultural crops	Metres
Apple	0.75–1.20
Apricot	0.65–1.30
Banana	0.30–0.60
Citrus	0.60–1.20
Grapes	0.45–0.90
Passionfruit	0.30–0.45
Peach	0.60–1.20
Pear	0.75–1.20
Strawberry	0.30–0.45
Vegetable crops	**Metres**
Bean	0.45–0.60
Cabbage	0.45–0.60
Carrot	0.45–0.60
Cauliflower	0.45–0.60
Cucumber	0.45–0.60
Lettuce	0.15–0.45
Pea	0.45–0.60
Potato	0.60–1.00
Tomato	0.60–1.20
Fodder and field crops	**Metres**
Cotton	0.60–2.00
Lucerne	0.60–2.00
Maize	0.60–0.95
Millet	0.30–0.60
Pasture	0.30–0.60
Soybean	0.45–0.75
Sugarcane	0.45–1.20
Sunflower	0.45–1.20
Wheat	0.75–1.00

Source: Cornish *et al* 1990

Some irrigators deliberately withhold irrigation at certain stages of growth to encourage root development.

The density of active roots is of some interest and importance. Rarely will roots be equally distributed throughout the root zone, but will tend to be more concentrated nearer the surface. It is no surprise that the top part of the root zone therefore dries more quickly. If the depth of the root zone was divided into four equal layers, an approximate estimate of the proportion of active roots in each layer could be 40:30:20:10 per cent, but this is highly variable between plant types, their growing conditions and their management.

Although it is desirable for soil water to be made available throughout the whole of the root zone, plants will survive quite happily if only part of their root zone is wetted, provided that part receives an adequate supply. The size of the root system is much larger than needed to meet current demand, and roots will become active if dry soil is re-wetted. Roots with access to plentiful soil water will work in preference to those facing dry soil. Although water will move toward roots, either under gravity or drawn by soil suction, roots can also grow toward water (horizontally as well as vertically). Although the density of roots may change with depth in the root zone, deep roots are as equally able to utilise soil water as shallow roots.

The above discussion has some implications to irrigation management:

- The need for more frequent irrigation during plant establishment and root development.
- The role irrigators can play in encouraging a large depth and volume of roots.
- During periods of low water availability, water can be targeted to part of the root zone.
- Under drip irrigation in more arid areas, root development may not become fully developed, unless system design and management are implemented correctly.
- If only part of the root zone is wetted (for example, some drip irrigation systems in more arid areas) irrigation will need to be more frequent, because the full crop requirement is being supplied over less than the full volume of the root zone.

Assessing requirements

Only a portion of the total amount of water held in the soil is readily available to plants. This is generally accepted as the difference in soil water contents between field capacity and wilting point, and is referred to as available water (AW). The AW varies with soil texture, according to Table 2.4.

Note that although these figures are derived from experimental results, they are likely to vary due to uncertainty in assessment of soil texture. Consequently, many AW estimates are quoted in a band or range.

For most agricultural projects, AW is determined indirectly by assessment of soil texture. It can be measured directly in a laboratory, by actually measuring soil suction at field capacity and wilting point.

Table 2.4 *Soil texture and available water*

Soil type	Available water at field capacity (mm per metre depth of soil)
Sand	50
Fine sand	75
Sandy loam	110
Fine sandy loam	40
Loam	165
Silt loam	175
Light clay loam	175
Clay loam	165
Heavy clay loam	225
Clay	140

Source: Cornish *et al* 1990

Irrigation water should be applied before the AW is consumed by the plant. Field capacity is often referred to as the optimum soil water content, since the plant is amply supplied with water, yet the root zone is not so wet as to retard plant growth and development. As the soil water content declines, the plant finds it increasingly difficult to extract water against soil suction, but this is particularly so after about 50 per cent of the AW has been consumed. This level of depletion is referred to as the refill point. Consequently, irrigation is timed to occur when no more than 50 per cent of AW, (the readily available water, RAW) has been transpired. The amount applied is sufficient to bring the soil water content back up to field capacity.

Example

Soil type	clay loam
Available water	165 mm/m
	(from Table 2.4)
Expected root depth	0.8 m (from Table 2.3)
Assume allowable depletion	50 per cent = 0.5
Calculate depth of irrigation	$0.5 \times 165 \times 0.8$
	= 66 mm

This example, gives the depth of irrigation to be applied to bring the soil water content back up to field capacity. It does not indicate when this will be needed, so it is necessary to measure or predict the rate at which soil water is being used (Chapter 3). Also, at the time of irrigation, the rate at which the water is applied (in mm/hour) needs to be matched to the intake rate at the soil surface. It also represents the amount of water to actually get into the root zone. Any water lost by direct evaporation, runoff or wind must be accounted for. To continue the example further, if this was a spray irrigation system with an application efficiency of 75 per cent, then the amount of water leaving the sprinklers would be equivalent to $66 \div 0.75 = 88$ mm.

The 50 per cent figure (the soil deficit) which defines the refill point is a management decision, which is selected to suit the situation, and varies according to a number of factors. Although it is preferred to apply water quite frequently, and therefore keep soil water content close to field capacity at all times, only microirrigation systems have this capacity. Spray and flood irrigation methods cannot be managed for frequent (daily) watering.

If the system is managed for 50 per cent depletion, then there will be little hope of the system coping with conditions that are worse than expected, and depletion levels may exceed 50 per cent before irrigation can be applied. For high value crops, or those particularly sensitive to water shortage at critical stages of growth, it may be more appropriate to have a target depletion of 20–40 per cent, but this may require greater capacity (and therefore cost) in equipment and earthworks. Also, if the system is managed so there is very little depletion of soil water allowed, then no advantage can be taken of natural rainfall if it is expected during the irrigation season.

Infiltration and water application rates

During moderate rainfall, or when irrigation water is applied, water will soak into the surface of the soil at a rate governed by the infiltration characteristics of the surface. A number of factors will influence the infiltration rate:

- The soil texture of the surface layer of soil, since this determines the particle and pore size distributions, which then determines the rate at which water can move between soil particles (the hydraulic conductivity of the soil).
- The soil structure and level of organic matter. (for example, poorly structured soils low in organic matter may develop a crust on the surface, inhibiting infiltration).
- The amount and type of vegetation cover.
- Whether the soil is already wet or dry. For example, a heavy soil which cracks open when dry will have a high initial water intake, but as the cracks close up when the

surface becomes wetter, the intake rate will fall significantly. The hydraulic conductivity of soil varies with its water content.

If the application rate of irrigation water (or rainfall) exceeds the infiltration rate, water will pond on the soil surface. Initially, this will fill the small depressions on the soil surface, but if continued, the depth of water on the surface will increase and runoff will occur.

For sprinkler irrigation, this would be undesirable, since the runoff of irrigation water is inefficient, and can contribute to undesirable wetting of the field. Table 2.5 provides a guide to the maximum application rate preferred under sprinkler irrigation.

For flood irrigation (such as border check or furrow irrigation methods), water is applied at the upper end of a graded bay or furrow, and applied at a rate which deliberately exceeds the infiltration rate. Water then flows down the gradient to irrigate the bay or furrow, so an "application rate" does not have the same meaning as in sprinkler irrigation. For drip irrigation, the application rates are (or should be) sufficiently low that runoff does not occur.

Under moderate to heavy rainfall, runoff from the irrigation area will occur. Sometimes this can be collected and stored on farm for use as irrigation water, particularly where the irrigation layout has been designed with a tailwater recirculation system.

Irrigation of heavy soils with low infiltration rates can sometimes prove awkward to manage, since a low application rate and a longer time of contact between soil and water will be required. This may appear to be a disadvantage, since

Table 2.5 *Preferred maximum application rates under spray irrigation*

Soil groups on texture and profile	mm per hour		
	Level to undulating	Undulating to low hills	Low to steep hills
1. Sands and light sandy loams uniform in texture to 2 m	31	25	20
2. Sandy loams overlying a heavier subsoil	20	16	12
3. Meduim loams to sandy clays over a heavier subsoil	16	12	10
4. Clay loams over a clay subsoil	12	10	7
5. Silt loams and silt clays	10	7	5
6. Clays	6	5	4

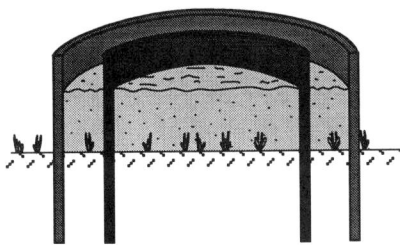

Figure 2.7 *A ring infiltrometer is an approximate but useful method of measuring infiltration rates in the field. It can also be used to approximately measure vertical hydraulic conductivity of soil layers other than at the surface. Two rings are tapped into the soil. Both are filled with water, but the rate at which water falls in the inner ring is measured*

more time is required to apply a given depth of water, but provided the selection of equipment and irrigation layout includes consideration of infiltration rate, an appropriate management scheme can usually be adopted.

In the field, infiltration rate can be measured using a ring infiltrometer.

Water movement through the profile

During irrigation or moderate rainfall on dry soil, the top layer of soil becomes saturated; that is, the pore spaces between soil particles are filled with water. Water will move downwards under the action of gravity, at a rate determined by the hydraulic conductivity of the soil, which is partly determined by the soil texture.

There is a distinct division between the saturated soil and the drier soil below it, and as water continues to be applied at the surface, the wetting front proceeds downward. If water application ceases, the wetting front continues downward, until the soil above the wetting front is at field capacity. There will still be a division between wet soil and drier soil, which will become less distinct over time as moisture moves through the soil in response to differences in soil suction, but these occur much more slowly than movement of gravitational water.

If water application is continued, gravitational water will continue downward until prevented by an impermeable layer or the watertable. If water application continues, the level of the watertable will rise upwards.

Evaporation will remove water from the top layer of soil, and plant roots will remove water from throughout the depth of the root zone. These will result in pores losing water, at a rate determined by a variety of factors (discussed further in Chapter 3). Water will move from pore to pore by capillary action (caused by surface tension forces) resulting in a slow redistribution of water from wetter parts of the profile to drier parts. In effect, drier soil has a higher soil suction, and draws water from wetter soil with a lower soil suction. Such movement can occur in all directions, and helps even out differences in water content that may result from non-uniform application of irrigation water. It also helps water migrate into beds in furrow irrigation.

CHAPTER 3

Evapotranspiration

S olar radiation provides the energy to vaporise surface soil water directly (evaporation E), and water contained in the foliage of plants (transpiration T). Water transpired from the foliage is replaced by soil water extracted by the roots. The combined effects of evaporation and transpiration are referred to as evapotranspiration (ET).

The rate of ET determines how quickly soil water is consumed and water consumed be ET is replaced by irrigation water (or rainfall). It therefore plays an important part in irrigation design and management, since it determines the interval between irrigations.

This chapter concentrates mostly on transpiration. Evaporation is significant in removing water only from the top few centimetres of bare soil, or on water lying freely on the surface of plants or the soil. Losses due to evaporation during irrigation are accounted for by applying an efficiency factor to the irrigation amount. Bare soil evaporation is of greatest concern during the initial establishment of crops, before a mature canopy develops. Under these circumstances, a different management approach is required anyhow, since a mature root system has not developed either.

Although listed here as separate processes, they are both driven by solar energy. It is usual to estimate the rate of transpiration by relating it to the rate of evaporation that would occur under the same conditions.

The transpiration process

The cell structure at the surface of plant foliage contains many cavities, separated from the atmosphere by a minute pore (stomate). The gas in the cavity is saturated with water vapour, extracted from adjoining cells, which creates a vapour pressure inside the cavity. The atmosphere outside the foliage is not

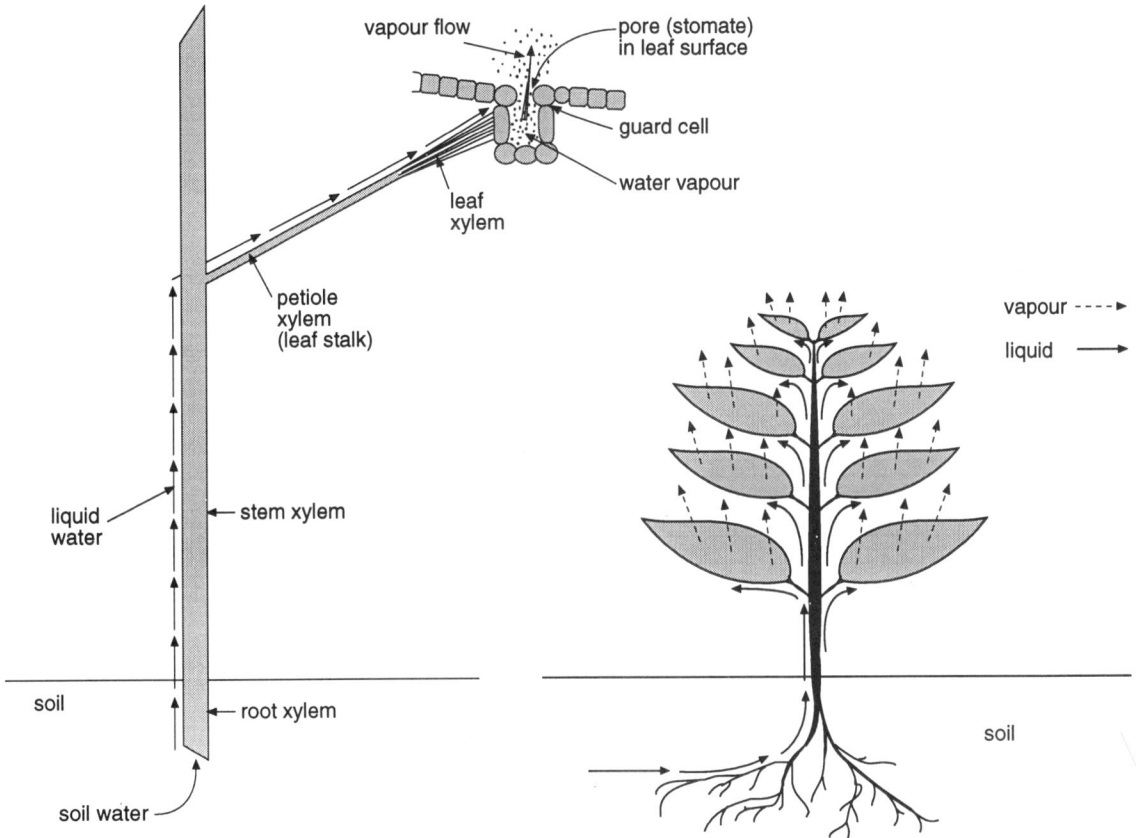

Figure 3.1 *Schematic representation of water flow through the plant. If water is readily avialable, and the plant healthy the rate of movement is determined by the relative vapour pressure of the atmosphere surrounding the foliage. (Source: Kelleher, 1981.)*

usually saturated with water vapour (100 per cent relative humidity, is rare), and so the vapour pressure of the atmosphere will be less than that in the foliage. Consequently, vapour moves through the stomata into the atmosphere, at a rate proportional to the difference in vapour pressures (the vapour pressure gradient) provided the plant is healthy and adequately supplied with water (that is it can transpire at its potential rate).

Consequently, the vapour pressure of the atmosphere surrounding the microenvironment of the foliage will determine the potential rate of transpiration, and this is influenced by solar radiation level, temperature, wind speed, and relative humidity. These variables determine the drying power or evaporative demand of the atmosphere and are highly variable in time and location.

As water vapour leaves the stomate, it is replaced by water contained in the foliage cells surrounding it, by the vapour pressure deficit created by transpiration, and by the process of osmosis, created by the presence of solutes in the cell contents. These cells are connected to fine tubes (xylem) in the foliage, which themselves are connected to more substantial xylem in the stems, then branching again in the root system. These are the conducting vessels that deliver water from the root system to the foliage. Water can move upward through them, against the force of gravity, due to the suction created by the vapour pressure deficit between foliage and atmosphere. Similar mechanisms allow water to enter the plant via root hairs, which are the interface between the plant and soil water.

Over 90 per cent of the water transpired by plants is evaporated into the atmosphere. The large flow of water is to bring dissolved nutrients to the plant, and to keep the plant cool.

Wilting

The structure of plant foliage is partly determined by the liquid pressure in the plant cells. If the plant is adequately supplied with water, and evaporative demand is not severe, the movement of water through the plant can usually keep up with evaporative demand, the liquid contents of the cells remains adequate, and the plant appears erect and upright (turgid).

Each stomate has a pair of special purpose cells guarding its opening to the atmosphere. In a turgid leaf, the guard cells keep the stomate open due to their particular shape. If evaporative demand exceeds the water uptake rate, water is lost from the foliage cells, including the stomate guard cells, which become flaccid and close off the stomate, reducing the rate of water loss. This gives the plant its typical wilted appearance, but also gives the plant a mechanism to control water loss.

Wilting can occur in plants adequately supplied with water, if the evaporative demand of the atmosphere is severe, exceeding the rate at which water can move through the root system of the plant. This will be a temporary condition, and the plant will restore turgor once the severe conditions have passed (usually overnight). Although affecting plant growth and development (stomatal closure also reduces gas exchange and therefore photosynthesis), there is not much that can be done about it, since soil water content is not the limiting factor. Canopy temperature can be reduced by sprinkler irrigation during this period, but water losses will be high, and a portable system will not be able to cover sufficient area.

If soil water content is low, wilting can occur even under moderate evaporative demand. If irrigation is not timed to replenish soil water, the plant will suffer a major setback in growth and development. The irrigation manager needs to

Figure 3.2 *Artificial windbreaks in this drip irrigated vineyard reduce plant stress, mechanical injury and wind-borne dust*

predict when irrigation will be needed, to avoid this condition. Other management factors could be considered:

- Selection of variety and timing of planting to avoid stress during expected periods of high evaporative demand.
- Provision of wind shelter to crops and orchards.
- Quick establishment of an extensive root system due to health and nutrition of the plant.

Estimating evapotranspiration

There are two common methods used to estimate the rate of ET in the field:

1. Measuring the rate of evaporation from a standard evaporation pan, and relating this to the ET of the crop.
2. Measuring the climatic factors which determine the rate of ET (radiation, temperature, wind, relative humidity) and calculating the rate of ET.

These are discussed in more detail later in this section. There are also other methods. A water budget can be drawn up, whereby changes to soil water content are measured frequently. If no water is added to the system (from rain or irrigation), it is reasonably safe to assume that water loss is due to ET, which enables it to be calculated.

Sophisticated instruments are available to measure:

- Sap flow through the plant.
- The vapour pressure within a leaf.
- The change in physical size of plant parts as they respond to moisture or moisture stress.
- The wavelength of radiation reflected from the canopy.
- The temperature of the canopy.

Figure 3.3 *Sophisticated scientific instruments, such as this phytomonitoring equipment, are now being applied to field management of irrigated crops*

These are commonly used in scientific studies of plant water use, but are now being developed for use in the field. They are indicators of the actual moisture status in the plant, rather than using indirect estimates derived from soil water or climate measurements. Experienced irrigators can observe subtle changes in the appearance of the crop, without taking measurements.

Pan evaporation

The potential rate of crop transpiration can be related to the rate of evaporation from a water surface. A standard method of measuring this is with an evaporation pan. The dimensions and construction of the pan are standardised and the Australian version of the pan has been in use in meteorological stations and on farms for many years. The water level in the pan is topped up each morning; the amount of water required to do this is a measure of the evaporation during the previous day. The amount of crop evapotranspiration (ETcrop) is related to the amount of pan evaporation (Epan) by the use of a crop factor (Kp), derived from experiments in the field.

$$ETcrop = Kp \times Epan$$

Although widely practiced, this method has some potential weakness, since the measurement of Epan is not very reliable, and the values of Kp not always specific to the crop or location in question. For example:

- A water surface is not the same as a canopy surface. It has different reflectance, surface roughness, and other physical differences, and behaves differently in the wind.

- Because of subtle differences in microclimate, within and between fields, Epan is very specific to the location of the pan (preferably within the crop).
- Epan varies depending on whether its immediate surroundings are dry or transpiring, bare earth or vegetation, and the extent to which these conditions extend around the pan. Hot dry air passing over an irrigated site will bring thermal energy with it, increasing ET.
- Birds and insects can interfere with readings, so it needs to be noted (and readings adjusted) if it is covered with netting.
- The system relies on human judgement to take the reading accurately, and record it accurately.
- There are different types of pan, and the readings need to be adjusted to suit the pan (different pans are adjusted by a pan factor) otherwise the crop factor is incorrect.

Unfortunately, pan data has proven to be a little unreliable, so it should not be used on its own. Pan data could be useful if the pan location is acceptable, if other weather or soil water data is used as well as pan data, and if the data is used for intervals longer than about ten days (over which the possible errors tend to be averaged out).

Calculated evapotranspiration

In this method, a weather station is used to measure solar radiation, wind run, temperature and relative humidity, which are the climatic factors which determine ETcrop. Rainfall is also recorded. Automatic weather stations are now readily available for location on farm, at reasonable cost. These are necessary to provide the site specific data that is required to give accurate readings. The data is logged at relatively short intervals, so that information can be derived for time periods of hours (even minutes) rather than daily. The data is downloaded into a computer, and evapotranspiration is calculated from a mathematical formula. Regional information can be used on individual farms after a period of calibration and experience indicate the differences between the two.

The calculation gives an estimate of ET by a "standard" or reference crop, (ETo), because different crops transpire at different rates depending on their physiology, stage of growth, and other factors. The reference crop is an actively growing green grass cover, of uniform height (8–15 cm tall), completely covering the ground surface, and adequately supplied with water. It is also assumed that the area of the reference crop is large enough to eliminate variations due to edge effects. Only in this way can the formula be relevant in different situations.

Figure 3.4 *On-farm weather stations are increasingly popular. Information can be integrated into irrigation scheduling decisions, as well as other management aids such as disease forecasting*

Consequently, the calculated readings of ETo need to be adjusted to suit the crop being grown, and its growth habits. The value of ETo is modified by the use of a crop coefficient, (Kc):

$$ETcrop = Kc \times ETo$$

As a final complication, some people use evaporation of an open water surface, (Eo), as their reference for potential crop water use. This would require a different set of crop related coefficients.

Crop coefficients

The value of Kc for any particular crop is determined experimentally, and is normally available from agronomy and irrigation advisers, or published literature. It may be difficult to find for unusual crops, where experiments have not been conducted, or for crops new to a particular district. In these cases, it is necessary to assume values of Kc based on crops with similar characteristics or water use, modified according to the environment of the site.

A significant influence over the value of Kc is the percentage of the ground surface that is shaded by the crop canopy. For annual crops, and deciduous perennial crops, Kc will be very small early in the season, influenced more by evaporation from wet bare soil or a cover crop than by crop transpiration. The value of Kc will increase during the season, being a maximum when the canopy is fully developed. The value of Kc will be higher for crops with a high plant density, for a similar reason. Note that ground cover is referring to

Table 3.1 *A sample of the crop coefficients available from advisors; in this case for own-rooted sultana grapevines under drip irrigation*

Month	Crop coefficient
September	0.10
October	0.23
November	0.30
December	0.40
January	0.40
February	0.40
March	0.30
April	0.23
May	0.17

Source: Drip Irrigation — A Grapegrowers Guide, NSW
Department of Agriculture 1993

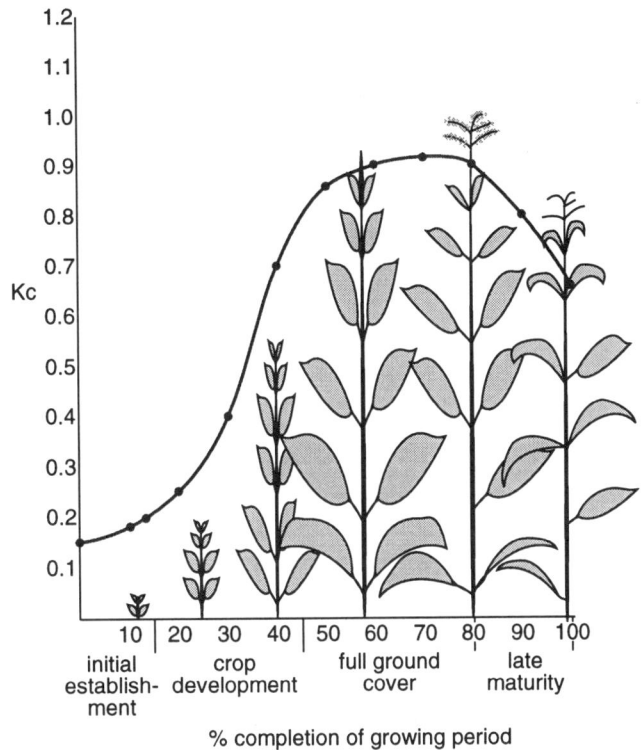

Figure 3.5 *Possible variation in crop coefficient over the growing season*

the percentage of the ground surface shaded by the canopy, not necessarily the highest volume of canopy itself.

Figure 3.5 shows a typical scenario for a crop such as maize. Other annual crops show a similar pattern, with differences in the exact shape of the curve.

Other factors influence the value of Kc in horticultural applications:

- The presence of a cover crop in an orchard or vineyard will tend to increase the value of Kc, since the cover crop will tend to compete for soil water, compared to evaporation from bare soil.
- Evaporation will be a more significant component of ET if frequent irrigations are practiced, wetting the surface of the site more often during the season, and this will tend to increase Kc.
- Localised irrigation which confines water application to the crop row will tend to decrease the value of Kc.
- The vigour of the crop, partly due to management but partly due to variety and rootstock selections.

It should be noted that Kc should be selected for each particular situation, and the values shown in this section are samples only. Similar comments apply to the selection of Kp,

Table 3.2 *Soil moisture factors*

Percentage of available moisture content actually present in soil		Range of free water evaporation Eo in mm per day			
		0–3.0	3.1–4.0	4.1–6.0	6.1 and above
A	100–75	1.0	1.0	1.0	1.0
B	74–50	1.0	1.0	0.8	0.6
C	49–25	1.0	0.7	0.5	0.35
D	24–00	0.5	0.3	0.25	0.15

Source: Garzoli, (1978)

for use with Epan. When selecting values for Kc or Kp, monitor soil and/or crop water status at least for a few seasons, to fine tune irrigation management. Take care to ensure that the Kc or Kp values you select are appropriate for the method used to measure transpiration.

The discussion so far has assumed the crop is adequately supplied with water. If not, then the actual rate of ETcrop will fall below the potential rate dictated by climate.

Actual versus potential ETcrop

The actual rate of ETcrop will fall below the potential rate because soil water becomes limiting, particularly during high evaporative demand. Table 3.2 lists possible values of a soil moisture factor, which is the ratio of actual to potential water use, depending on the percentage of available water in the soil and the rate of evaporation. (It is assumed that the plant is free of disease and adequately supplied with nutrient.)

Peak irrigation demand

It is important to estimate the maximum likely amount of water the crop will require. More specifically, the maximum rate at which soil water could be consumed, since soil and crop characteristics determine the volume of water that must be applied, and must also be calculated. This will determine the maximum capacity of the irrigation system, which in turn dictates the size of reticulation and pumping equipment. It will occur when evaporative demand is at its greatest, the crop has a fully developed canopy and root system, and is transpiring at its potential rate (that is adequately supplied with water, and in good health and fertility). Maximum likely requirement is the most difficult to determine, requiring a weather prediction, based on historical records.

The usual approach is to analyse rainfall and pan evaporation data for the locality. Rainfall data is readily available, and already processed in a form which is useable (in particular, average monthly rainfall, as well as the probability of receiving high and low rainfall amounts in any month). Long duration records enable reasonably high confidence in these figures.

Evaporation data needs to be analysed in the same way. However, there are fewer localities with reliable evaporation records, and only some of these have a sufficiently long duration of recording to give confidence in their statistical analysis. Average monthly pan evaporation is reasonably available in tabulated or map format.

Analysis of average monthly pan evaporation data has some limitations when used to predict evaporative demand and therefore irrigation requirements:

- Pan evaporation data is not very reliable, and may not truly represent potential transpiration in the field in question.
- Short duration records do not give a reliable indication of conditions that might occur in the future.
- Average monthly figures do not indicate the range in extreme conditions that might be experienced (these are the important ones in estimating peak water demand).
- Average monthly figures smooth out the peaks and troughs in evaporation that occur during time periods of lesser duration.

Despite these limitations, average monthly rainfall and evaporation data is used to give an indication of water demand on a seasonal basis, and a sample calculation is shown in Table 3.3 for a range of rainfall scenarios.

It is necessary to apply some judgement to the figures such an analysis might provide. The estimate of water requirement should be increased if the value of the crop is high, the risk of yield loss from a dry spell is high if it is a perennial crop or plantation, or the irrigator has a conservative approach to risk.

The interval between irrigations may be much less than a month (as short as a day or two for some microirrigation systems) where evaporation may be much higher than the monthly average may suggest.

Consequently peak water demand based on average monthly data will be seriously underestimated. If historical records of evaporation are available for the locality, or there has been sufficient experience in growing the crop in the district, a better estimate can be made. The full evaporation

Table 3.3 *Sample calculation of seasonal water requirements*

Location:	Canowindra, NSW
Crop:	winegrapes
Vine spacing:	2 m
Row spacing:	3 m
Area per vine, A:	$3 \times 2 = 6$ m^2

	Rainfall, R. (mm)		Average evaporation	Crop factor	Water requirement (L /vine) (Epan. Kp – R) × A		
	Average	Median	(mm) Epan	Kp	Average rain	Median rain	No rain
Oct.	58	49	159	0.15	nil	nil	143
Nov.	50	41	209	0.25	14	68	313
Dec.	52	41	280	0.3	192	258	504
Jan.	61	47	272	0.4	288	371	653
Feb.	49	31	218	0.4	229	337	523
March	45	32	189	0.3	70	148	340
April	44	36	126	0.25	nil	nil	189
Total (L/vine)					793	1182	2665
Total (ML/ha)					1.32	1.97	4.44

Notes:

1. Rainfall and evaporation records from the Bureau of Meteorology, Station no. 065006.
2. Crop factors from published data.
3. Nil water requirement noted when rain exceeds ETcrop.

The results clearly show the influence of rainfall during the season on crop water requirements. To expect average rainfall is too risky for water supply planning. No rain during the season is a worst-case scenario.

record can be interrogated to determine the variation in data from the average, in an attempt to determine the frequency and extent of extreme conditions.

IRRIGATION PLANNING IN A VARIABLE CLIMATE

The following analysis uses weather records (1976–1996) from Orange (Agricultural Research and Veterinary Centre) as a sample indication of variability. Note that Orange has a cool climate, and is relatively reliable, so the conclusions may be more significant at other localities.

At Orange, the greatest evaporative demand occurs, on average, in December and January.

Average daily December Epan 7.1 mm/day
Highest December Epan 10.0 mm/day
(daily average for the month)   (in 1979)
Lowest December Epan 3.7 mm/day
(daily average for the month)   (in 1995)

The actual evaporation record shows that during December 1979, three consecutive days recorded 11.0, 13.4, and 12.4 mm evaporation, which has an average daily evaporation of 12.3 mm/day. Three days could coincide with the irrigation interval for some shallow-rooted horticultural crops, where peak daily evaporation exceeds the average by 73 per cent. An irrigation system designed to cope with Epan of 7.1 mm/day will not cope with such extreme conditions.

Average December rainfall is 90 mm (median is 57 mm) so you could expect some of the crop's water requirements to be met by this, depending on its effectiveness.

The analysis is more complicated because in December 1979 only 0.4 mm rainfall was recorded (none of which would be effective). Luckily, the previous month recorded 44.4 mm and the following month 57.7, so it is unlikely there would be a severe crisis.

This may not have been the case in 1982/83, when the following rain was recorded:

October	12.6
November	7.0
December	8.6
January	49.8
February	28.3
March	78.2
Season Total	184.5

The total recorded for 1982 was only 323.9 (average is 933) so the irrigation season would have commenced with very low subsoil moisture levels, and the crop would be highly dependant on irrigation. Luckily, April rain of 97.4 mm heralded a welcome Autumn break.

CHAPTER 4

Irrigation scheduling

Chapter 2 described a number of techniques for measuring the status of soil water (water content or soil suction, relative to field capacity) and how this helps decide how much water to apply. Chapter 3 described methods for estimating the rate of evapotranspiration. Measurement of soil water status and its actual rate of depletion can help predict when the next irrigation is required. Irrigation scheduling is the use of this information to decide when to irrigate.

Water budget

Figure I.1 shows how water is added or lost from the root zone. Water budgeting is a simple system of measuring and accounting for each component, usually on a daily time step. If the water content is known at the start of a period, it can therefore be monitored on a continuous basis. For many situations, minor factors can be assumed to be small (although there are some exceptions) so that additions to soil water content are attributed to rainfall and irrigation, and subtractions by evapotranspiration. Figure 4.1 shows a hypothetical water budget in its simplest form.

In Figure 4.1, soil water content was at field capacity on day 1, but ETcrop reduced soil water content progressively, until the refill point was reached. Irrigation during the next day replenished soil water back to field capacity. This technique is extended over the growing season, and allowances are made for effective rainfall. Experienced irrigators, with their eye on the weather map, can calculate the water budget into the future, and therefore predict when the next irrigation is likely. Unless water is available on demand, this will be necessary, since:

- Water may need to be ordered some days in advance of requirements.

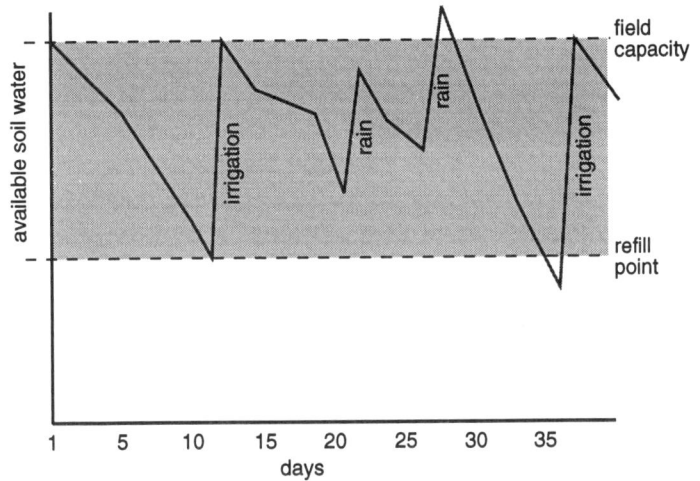

Figure 4.1 *A water budget is a simple method to monitor soil water content. It can account for the contribution of effective rainfall to soil water, and can help predict when irrigation is next required.*

- During peak irrigation periods, with portable or shifting irrigation systems (that includes most spray and surface irrigation methods), equipment may still be committed to completing the previous irrigation cycle.

A long range prediction can be based on historical records of water use at the location.

Water budgeting can be conducted using direct measurements of soil water status, or by estimates of ETcrop based on published or measured values and local crop factors. However, given the difficulties in accurately measuring these factors, it is preferable to use more than one measuring technique so that readings can be cross-checked.

Example

Continuing the example from page 46:

Depth of irrigation	66 mm
Average Epan for the period	8.6 mm/day (see note below)
Crop factor for the same period, Kp	= 0.65
ETcrop	= 8.6 × 0.65
	= 5.6 mm/day
Therefore expected irrigation interval	= 66 ÷ 5.6 = 11.8 days

Note that this example uses an estimate of Epan, which has some limitations compared to calculated ETo. The crop factor is selected from available information, but factors used with Epan will be different from those used with ETo. Measured data can be used to monitor current water use, but when trying to predict future water use, it is assumed that weather over the next few days will be similar to that over the last few days. For long term planning, historical records of evaporation can be investigated to give estimates of average or peak water consumption for the period in question.

The calculation shown in the example assumes that the canopy and root system extend evenly across the ground surface, and that water is uniformly available across the full area. This is not necessarily the case under microirrigation of row crops, where water is applied to a strip under each crop row, and adjacent wetted strips might be separated by a strip of dry soil. In this situation, ET is assumed to occur at its potential rate, since the plant canopy is likely to utilise most of the available solar radiation (that is, there is close to full shading of the ground surface) but the supply of water is limited to a portion of the available soil volume.

As an approximation, if the wetted soil volume is half of the total volume, it still needs to supply the full water requirement of the crop, and so the available water will be consumed twice as quickly. Hence, the irrigation interval will be half that which would be calculated by the method in the example.

A further weakness of the method presented in Figure 4.1 is that it does not show the variation in soil water status with depth through the root zone. Hence, it may give a misleading picture of soil water availability to the crop. Figure 4.2

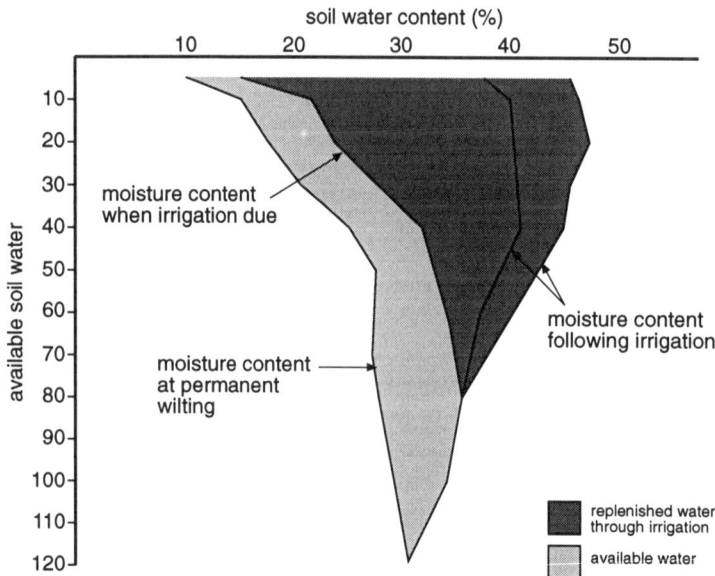

Figure 4.2 *Typical soil water information from a neutron probe (Source: Browne, 1984)*

Figure 4.3 *Typical EnviroSCAN® output. Soil water monitoring data of onions plotted as cumulative soil water content in the 10, 20 and 30 cm layers. The crop was grown under a centre pivot irrigation system on a sandy loam in South Australia. After the last irrigation event on the 16 February (A) the crop was allowed to dry out for harvesting. A steep slope (B) indicating unrestricted crop water use (17–19 February) is replaced by significant changes in the rate of daily crop water use (C) on the 20 of February, indicating the onset of crop water stress, which is used to set an irrigation Refill Point. Diurnal fluctuation of soil water beco 1mes evident under plant water stress (from 21/2-5/3). (Source: Buss, P. [undated], The use of capatitance based measurements of real time soil water profile dynamics for irrigation scheduling. Sentek P/L, SA.)*

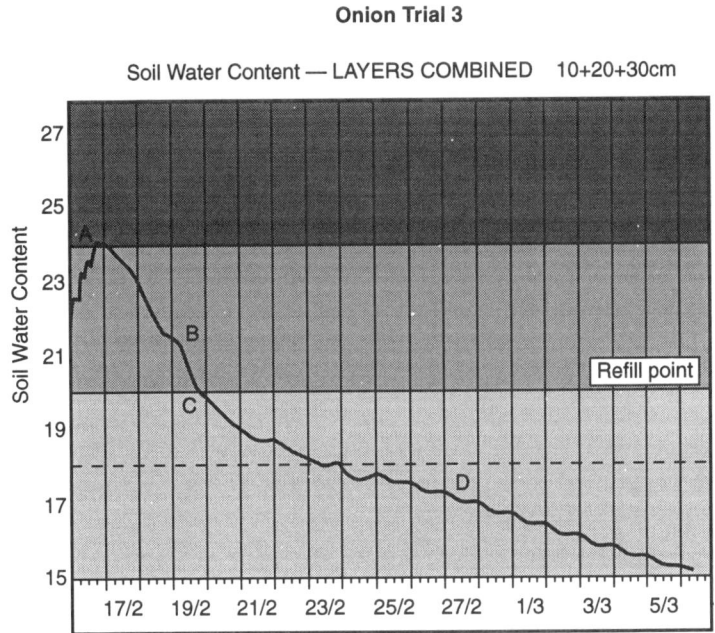

Onion Trial 3

Soil Water Content — LAYERS COMBINED   10+20+30cm

represents the type of information available from a neutron probe, where soil water status is monitored at numerous depths at each measurement site.

This type of information can greatly assist in more accurate timing and application amounts of irrigation and can help detect overwatering (and subsequent deep percolation) and underwatering (where a wetting front does not proceed sufficiently through the profile). It can also assist detection of other problems, such as poor subsoil or fertility conditions restricting root development, reflected in less water extraction than expected at those depths.

Figure 4.3 shows the typical output from a real time sensor such as EnviroSCAN®. In addition to information at various depths, continuous monitoring instruments such as these can detect quite subtle changes in water use over short time intervals.

Allowance for rainfall

Not all of the rain measured in the farm rain gauge becomes available to plants. During high intensity rain, some may run off the irrigation area, especially if the profile is already full.

If heavy rain is forecast, it may be appropriate to delay full irrigation, to take maximum advantage of rainwater which does enter the profile.

Very light rainfall amounts (say less than 5 mm) are usually ignored during irrigation scheduling, since most is likely to evaporate from wetted soil and foliage, and little is likely to infiltrate.

Moderate rainfall amounts entering a profile at a water content less than field capacity can be "credited" in the budget, enabling irrigation to be postponed. Unfortunately, if the profile was approaching the refill point prior to rain, the amount of rainwater may not be sufficient to replenish the full profile. Rainwater will create a wetting front which will proceed downwards until infiltration of rainwater ceases and field capacity is reached. This may result in a band of wetted soil above a band of dry soil. Roots will preferentially take water from the band of wetter soil, which may therefore be consumed faster than the average water content may suggest.

Careful monitoring of soil water status at various depths is the best way to check how effective rainfall has been.

Gross irrigation application

Some water is lost between the point of water supply and the arrival of water in the root zone of the crop. The amount of water lost varies widely between different reticulation and application methods, and for different soil types, irrigation layouts, and so on. These losses are primarily due to evaporation, seepage, wind (in the case of spray irrigation) and runoff.

Water loss before arrival at the root zone results in an inefficient application, and is usually expressed as a percentage of the water delivered. The gross amount of water delivered from the supply must be greater than the crop water requirement by this amount. Inefficiencies can be minimised by good irrigation design and layout, and good management practices.

Deep percolation below the root zone will also result in inefficient water use, and should be avoided if possible. However, some irrigators, particularly in salt affected areas, find it necessary to apply an excess amount of water to leach accumulated salts from the root zone. The amount of water required for leaching will depend on the salinity levels of the soil water and the irrigation water, and the salt tolerance of the crop, and is also expressed as a percentage of the crop water requirement. Between five and ten per cent is common. It is also necessary to check conditions below the root zone, since it will be highly undesirable to leach salts into a natural drainage area, or if the watertable is close to the root zone.

Example

Net application (from previous calculation)	66 mm
Leaching requirement 5 %	3.3 mm, say 4 mm
Total	70 mm
Application efficiency	80 % = 0.8
Gross application	70 ÷ 0.8 = 87.5 mm

Deficit irrigation

Some irrigators apply less than the calculated crop water requirement on occasions to manipulate the growing conditions of the plant. For example, allowing soil water to be depleted to less than the normal refill point may encourage development of a more extensive root system, as roots will extend toward sub-soil water. A larger root system is often preferred, since it has access to a larger reserve of soil water, reducing dependence on more frequent irrigations. It may also increase anchorage of tree crops against strong wind.

Regulated deficit irrigation has been used with some fruit crops to reduce vegetative vigour. This is done to control eventual fruit numbers and fruit size, to produce higher yields and returns. It may also result in less water use, less pruning and easier management of orchards with close tree spacings. Naturally, the timing, level and duration of the deficit period must be carefully managed to ensure water stress does not create adverse effects. Readers are referred to other texts for details on these techniques.

CHAPTER 5

Surface irrigation

Surface irrigation refers to methods of water application where a body of water, of some depth, is applied to one end of a bay or furrow. Because the bay or furrow is constructed with a slight gradient and because the body of water has a natural tendency to spread, water flows down the bay or furrow in a sheet or stream, infiltrating as it goes.

These methods are often referred to as flood irrigation, but that is not a good description for furrows or bays where reasonable control over the water movement can be created. They are also sometimes referred to as gravity irrigation methods since gravity is the driving force causing water flow, once water is delivered to the bay or furrow.

Much of Australian irrigation is performed with surface irrigation, but there are some distinct limitations with this method of watering.

Surface irrigation methods

A number of different versions of surface irrigation are practiced. Furrow irrigation is employed for irrigation of row crops. For field crops, border check irrigation is the most common, but other versions are also in use, their selection depending on the slope (or lack of it) at the site, the crop grown, the availability of water, and the degree of control required over water flow.

Furrow irrigation

Crops are established in raised beds, enabling water to flow down defined furrows. Successful irrigation requires water to soak laterally into the bed as well as downward into the root zone. Furrow spacing is therefore partly determined by the hydraulic conductivity of the soil and subsoil, but also by preferred crop row spacing and machinery requirements.

Figure 5.1 *Furrow irrigation of cotton. In this example, water is syphoned from the supply channel to the head of the bay.*

Water is admitted to the head of the furrow by a siphon in the case of water delivered in open channels. Low pressure gated pipe or fluming is also available, and this occupies much less land area than channels (some can also be buried). It is well suited to high value crops where land area may be limited, or where seepage from a channel would be excessive.

The length of furrow is determined by the slope along the furrow and soil infiltration characteristics, since these determine both the travel speed of the wetting front as it proceeds down the furrow (and therefore the time taken to reach the end of the furrow, and the time water is in contact with the soil) and the amount of water entering the profile. These are characteristics specific to each site. Other factors include the amount of earthmoving required to create the desirable slope for the particular soil type, the shape of the field, and a preference for uniform furrow lengths in each field to assist crop management.

Figure 5.2 Below, *Furrow irrigation of grapes, using above ground low-pressure delivery pipe.* Below right, *Furrow irrigation of citrus, from buried pipe and risers in this orchard.*

Furrows are normally constructed down the slope, but occasionally furrows are located on the contour to enable this method of irrigation on steeper slopes (usually for orchards and plantations).

For annual crops, the field is cultivated after harvest, and beds are reconstructed before the next planting. This can have long term detrimental effects on soil structure, organic matter levels, and activities of soil organisms. Some farms have established permanent beds; ie the beds and furrows are always retained. This requires a re-think of crop management techniques, and machinery requirements, but significant improvements in crop performance have been reported. Improvements are due not only to better soil quality in the beds, but also due to improved timeliness of operations.

Border check irrigation

This method is used extensively for a wide range of grain and fodder crops and pastures. A series of bays are formed within each field, running down the slope. The borders of each bay are formed by constructing small earthen check banks (hence border check). Water is admitted to the head of the bay using syphons, gates or pipes in or over the wall of the supply channel. It quickly spreads across the head of the bay, and a wetting front proceeds down the bay.

Like furrow irrigation, the length of the bay is determined according to the slope and soil infiltration characteristic. The cross slope of a bay is limited to 25 mm (preferably less) so this will partly determine the width of bay. Check banks are a possible nuisance to machine operations, so it is preferably to have bays as wide as possible. If they are too wide, uneven watering will result.

Naturally, the surface of the bay must be built and maintained at the correct, uniform gradient. Unevenness in the bay surface will create uneven watering and poor drainage, reducing crop performance. This is noticeable just after irrigation (or moderate rainfall) where water may pond on the lower parts of the bay for too long.

Where bay cross slope is a problem, the surface of the bay can be corrugated. This helps guide water down the bay, providing a more even watering, without interfering with crop management.

Where cross slopes are excessive, check banks can be constructed on the contour rather than down the slope, provided the vertical interval between adjacent banks is 25 mm or less. This will tend to give uneven watering, and the bay width is likely to be uneven. From a technical and management point

Figure 5.3 *Border check irrigation. In this example, water is admitted to the bay by a pipe in the wall of the supply channel.*

Figure 5.4 *Rice growing in a flooded basin.*

of view, it may be preferred to landscape the area to give uniform width and grades to the field.

Other versions of surface irrigation

On flat sites where drainage is difficult at the end of irrigation, the borders of irrigation bays can be formed by ditches instead of check banks.

For irrigation of rice, level bays are formed and flooded, water flowing continuously from bay to bay during the irrigation season. Hillside flooding is occasionally practiced for irrigation of pasture, where a ditch delivers water approximately on the contour above the irrigated area. A temporary check is placed in the channel, water builds up in the channel and floods over the side.

Characteristics of surface irrigation

Because the gradient of the land is an essential element of water flow and control, the land must be formed so as to give a precise gradient, predetermined according to the situation (Figure 5.5).

Consequently, large efforts are made to accurately survey the existing land surface, and move the topsoil until the required gradient is achieved. High points are cut away and that earth is moved to fill low points, using earthmoving scrapers, often guided by laser control systems. Efforts are also made to create a smooth surface, using large land planes (Figure 5.6). Bays or furrows are then constructed on this surface.

The resulting gradient partly determines the speed with which water gravitates from the top to the bottom of the bay or furrow. Consequently, the chosen gradient must not only

Figure 5.5 *Earthmoving under way using a tractor drawn scraper. The operating height of this is adjusted in response to a laser signal. The mast on the machine carries the laser receiver; the laser emitter is mounted on a tower near the centre of the field.*

Figure 5.6 *A land plane provides final smoothing to the land surface.*

suit the natural site conditions (you want to shift as little earth as possible) but must also be matched to the soil infiltration rate and the length of the bay or furrows.

Apart from the average gradient between the ends of the bay or furrow, it is important that there be little variation of gradient along the full length. Even minor variations will create depressions that collect water and interfere with uniform water flow. (The exception might be where the bay or furrow crosses different soil types.) Further, the gradient across the field must be considered, because excessive cross-slope will also interfere with uniform watering and could result in excessive earthmoving.

In developing a site for surface irrigation, extreme care must be taken not to cut too deeply, exposing subsoil during earthmoving operations. Not only will this require special treatment to establish crops, but the fill areas will settle below the cut areas, upsetting the gradient. It may be necessary to land plane the site for some seasons after initial construction.

Specialised services and machinery are required to survey, calculate depths of cut and fill, and perform the earthmoving operations.

Reticulation of water

Surface irrigation methods usually utilise open channels to distribute water about the farm. The main supply enters the property usually at the highest point, so that secondary channels reticulating from this point can do so using gravity. The ability of the channels to command higher ground is a major determinant of the final layout of the irrigation system. It is often necessary to build up the ground under proposed channels to provide this command. Channels are constructed using specialised earthmoving machinery and like the bays or furrows they supply, must be installed on a precise gradient.

In the same way that valves and fittings control water flow in pipes, there are various structures installed in channels:

- Gates to control the direction of flow.
- Checks and weirs to stop water flow and increase water level in the channel.
- Drop structures, where it is necessary to bring water to a channel lower in height.
- Culverts, where access roads are required to cross the channel.

Various means are used to deliver the water from the channel into the head of the bay or furrow.

- Syphons. The water level in the channel is higher than the water level on the bay or furrow, so it will easily syphon, using aluminium or plastic tubes. This system is common with furrow irrigation, where one siphon is used for each furrow. It is also used for bays. The siphon is started manually at the start of the irrigation, and the irrigation can be stopped by draining the channel or by removing the siphon. Using syphons has a high labour cost (except for the larger diameter syphons, shifted and started by machine).
- Gates. Used with bays, a gate is opened in the wall of the channel.
- Pipe outlets. Also used with bays, a pipe is installed in the wall of the channel at the time of construction. A cap or plug is removed to allow water from the channel.

Many versions of the above techniques are in use, in an attempt to maximise efficient use of labour.

Channels have some disadvantages:

- Construction cost.
- Seepage, leakage and evaporation of water, significantly reducing application efficiency.
- Large land areas must be devoted to them.

In more intensive horticultural applications, alternative distribution methods are common using low pressure pipelines. In the "gated pipe" system a pipeline is installed at the top of the field, with gates allocated to correspond with each furrow. Distribution losses are eliminated, and minimum land area wasted, but at the cost of the pipelines and pumping equipment necessary. The number of furrows that can be watered at any one time is limited by the flow capacity of the pipeline.

Various types of rigid and "lay-flat" plastic pipes are available as alternatives to channels in furrow irrigation layouts. In some applications, buried pipes with risers at each furrow head are used.

Management

Surface irrigation methods apply a relatively large amount of water at relatively infrequent intervals. It is difficult to apply a small amount of water with any evenness. Further, in most areas where surface irrigation methods are used, water is supplied from regulated sources and ordering of water in advance is required.

This has a number of implications. Because large amounts of water are applied at long intervals, some crops may not be ideally suited, as moisture stress may start before the next irrigation. Also, since large amounts are applied at the time of irrigation and because of the method with which the water is applied, temporary waterlogging occurs for a short period during and after irrigation. When water is in short supply (when allocations are reduced during drought), it is difficult to give precise control over a small depth of watering because surface irrigation methods require a certain depth of water to provide the driving force causing the water to flow.

Efficiency of watering

Consider a single bay or furrow, of a fixed length and gradient and with soil with a particular infiltration rate. At the time irrigation commences, water first contacts the soil at the head of the bay or furrow and starts infiltrating at that point. As more water is applied, the water spreads across the bay, builds up in depth and starts flowing down the bay or furrow. It will

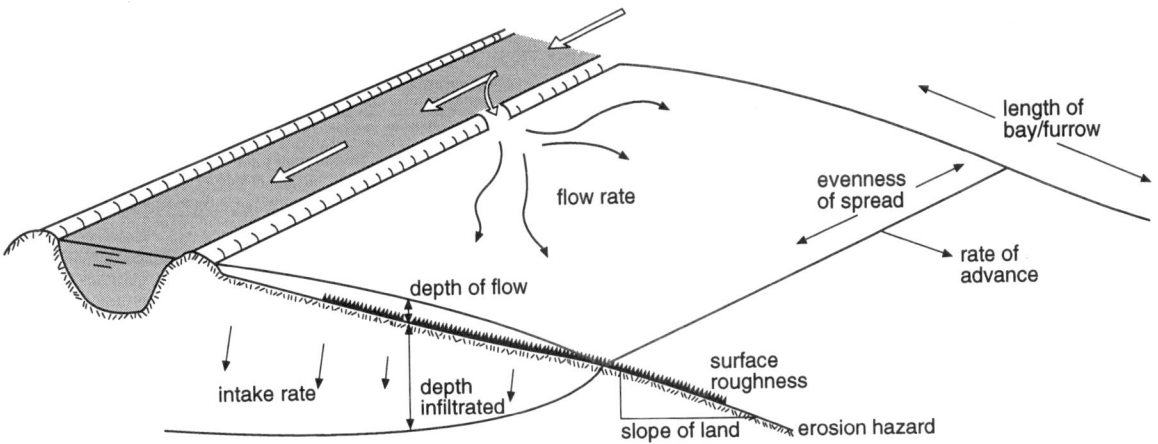

Figure 5.7 *A number of factors influence the amount of water entering the profile under surface irrigation, and the uniformity of that application.*

take some time before water reaches the end of the bay or furrow, depending on slope, length, flow rate and infiltration rate.

During that time, water has been infiltrating at the head of the bay or furrow, whilst nothing has reached the bottom. Consequently, surface irrigation methods have an inherent unevenness in the amount of water applied. If the water is applied too long at the top, water in excess of the crop's requirements will percolate to the watertable and be "wasted", during the time it takes to adequately water the lower end. If only enough is applied at the top, then the bottom end will not receive its requirement. Either way, water application efficiency, and potentially the yield, are affected.

This situation is complicated further by the fact that minor variations in gradient along or across the bay or furrow may result in localised ponding, and soil infiltration rates may vary across the field and with the amount of water received.

Various strategies can be employed to minimise these problems:

- Attention to accuracy during landscaping.
- Varying the gradient along the bay or furrow to control velocity of water flow (not easy to construct).
- Change the flow rate during irrigation. Start with a high flow rate to get the water to the end quickly, then cut back the flow rate so it just matches intake rate along the full length. This is also not easy to manage.
- Construct bay or furrow length to suit soil intake rate.

Where gated pipe is used, "surging" techniques are worth considering (although the cost of special valves must be considered). The area being irrigated is divided into two parts (left and right). A special valve supplying the gated pipe sends water one way initially at a higher than normal flow rate to send a pulse of water down the furrow. After a shorter than normal period of time, the valve switches the other way, to send a pulse down those furrows. This switching is repeated frequently during the irrigation.

Because of their characteristic inefficient application of water, surface irrigation methods often contribute to groundwater accession and rising watertables.

On-farm efficiency can be improved in another way. In order to ensure adequate water is applied at the downstream end of the bay or furrow, it is frequently the case that water runs off the end of the bay or furrow during irrigation. This water will be wasted and contributes to reduced application efficiency (except where it runs onto pasture for improved grazing production but this is not usually "planned").

Most new installations construct drains to collect this tailwater and deliver it to storage dams at the lowest point of the system. In some designs, and with careful attention to levels and gradients, the tailwater drain can be terminated adjacent to the main storage reservoirs with tailwater pumps used to transfer it to the main storage.

Such systems can also be used to capture storm runoff to reduce dependency on delivered water.

Labour

Surface irrigation methods are traditionally labour intensive, particularly associated with getting the water on to the field and checking when to terminate watering. Syphons are particularly time consuming.

Figure 5.8 *Tailwater is collected in a sump, and pumped into the adjacent storage dam.*

Figure 5.9 *In this layout, a single large diameter outlet pipe services many furrows. The hinged gate on the pipe can be opened and closed from a vehicle, saving time.*

Various devices are available, but not common, to minimise labour:

- "Giant" syphons, mechanically activated, to service multiple furrows from a head bay.
- Gates and other channel outlets, in preference to syphons.
- Various devices to automatically close channel checks, to start irrigation in a new part of the field.
- Sensors and transmitters to notify when irrigation water has progressed a certain distance down the bay or furrow.

Layout can also contribute to efficient labour. Long runs and wide bays mean fewer outlets per hectare, but going too long and wide may contribute to non-uniform application.

Summary of design procedures

An estimate is required of the expected water requirement (seasonal and peak) for the area proposed, and the water supply is investigated to ensure supply is matched to demand. Initial feasibility planning should identify the approximate boundary of the irrigated area, based on limitations created by soil characteristics or topography.

A detailed topographic survey is conducted to enable accurate measurement of the slope of the ground surface. Rarely will this be satisfactory without some earthmoving. The preferred slope along proposed bays or furrows is estimated, based on soil characteristics and desirable length of run. The difference between actual and proposed slopes enables a prediction of the amount of cut and fill required at any point of the area, and therefore of the total earthmoving requirements.

Figure 5.10 *One type of sensor and its receiver to detect progress of the wetting front during irrigation. (Courtesy: Electronic Irrigation Systems P/L)*

This can be calculated by special software directly from the survey data.

Adjustments can be made to account for various management factors, provided application uniformity is acceptable:

- Long runs, of equal length in each block, make it easier to manage during irrigation, and for most crop husbandry operations.
- Minimise the depth of cut, to reduce risk of topsoil removal..
- Large individual paddocks assist mechanisation.
- Layout should maximise the use of the irrigated area, with minimal "waste" areas in channels, drains, roads, etc.
- The number of structures and earthworks should be minimised.

Earthworks for final layout can then be costed, and pumping stations designed. The highest point to be irrigated is located. Because water is reticulated by gravity, usually through channels, this point will influence much of the proposed layout. The main supply channel is located to deliver water to irrigation blocks within the area, whilst being constructed at the correct gradient. In a similar way, tailwater drains are positioned to collect water from each block, and gravitate it to a sump at the lowest point of the layout.

Survey information is pegged onto the site, and earthmoving is conducted usually under laser control. The quality of the earthmoving job should be checked by comparing levels of the finished work with the levels proposed in the irrigation plan. Channels are constructed and structures installed.

Rubber-tyred elevating or carry scrapers are used where earth must be hauled from one point to another. Land planes are used to smooth the surface following scraping. This is performed in a number of compass directions to ensure the final surface has exactly the correct down-slope and cross-slope.

Bay checks, or furrows, are then constructed on the finished surface. Graded land will settle after irrigation, and re-grading may be required several times initially, and occasionally in subsequent years.

CHAPTER 6

Spray irrigation

S pray irrigation is an attempt to simulate natural rainfall by creating and spreading droplets over the field. Droplets are generated by the breakup of a stream of water exiting a nozzle, caused by pressurising the water in the pipeline leading to the nozzle. The trajectory from the nozzle and the rotation of the nozzle distribute the droplets over a circular area. Consequently, spray irrigation systems are characterised by:

- A sprinkler which houses the nozzle(s) and provides a mechanism for rotation.
- A pipeline network to convey water from the water storage to the sprinkler(s).
- A pump and power source to pressurise and reticulate the water.

The design of a spray irrigation system therefore requires considerable attention to the performance specifications of each component and ensuring each component is integrated into the system. Further, pumping equipment and pipelines have quite specific design limits; that is their optimum performance is usually contained within a specific range. Consequently, it is important to consider the design capacity of the system and the expected management plan for the crop in the early planning stages.

Suitability of spray irrigation

Methods of spray irrigation vary substantially, but the typical applications for spray irrigation, and its advantages and disadvantages, follow:

Topography

Spray irrigation can be successfully used on sites where excessive earthmoving would be required to initiate a surface irrigation system. On sloping sites, however, runoff can be a problem if the spray application rate exceeds the infiltration rate of the soil. On steeper sites, additional problems of pressure variation and the tracking of self-propelled machines can arise.

Soil conditions

Application rates from sprinklers can be more closely matched to soil infiltration rates by appropriate selection of nozzle sizes and operating pressures. This makes spray irrigation more suited to soil types where application by surface irrigation becomes inefficient such as on light soils or very heavy soils.

Shallow soils where even moderate earthmoving for surface irrigation may expose the subsoil, may be irrigated with low application sprinklers.

Where crusting of soil is a problem, impact by droplets on bare soil may exacerbate the condition.

Water supply

Spray irrigation may make better use of limited water supplies, compared to surface irrigation, because of a higher degree of control over volume and depths of water applied. (This advantage is even greater with microirrigation methods.) Such control makes spray irrigation more suitable for germination and emergence of seeds and seedlings, provided crusting created by droplet impact is avoided. Distribution and reticulation losses are minimised in a piped system, compared to the open channels typical for surface irrigation. On the other hand, wind drift of spray droplets can severely interrupt the uniformity of application from sprinklers and reduce application efficiency.

Sprinklers apply water typically over the whole crop canopy, which will increase canopy humidity, and may cause problems of accelerated fungal attack and deposition of any dissolved salts onto foliage. However, spraying irrigation water over the crop canopy may provide secondary benefits of dust control and cooling benefits.

System management

Manual shifting of spraylines or irrigation machines often contributes a significant portion of the cost of operation. Labour can be reduced with self-propelled machines, which are more easily managed than a channel/flood system. The equipment must remain operational in wet and/or cultivated soil.

The design capacity of a spray irrigation system sets a limit for the application rate and turnaround time of the equipment. Consequently some spray irrigation systems have difficulty coping with extreme conditions of water demand, particularly when strong winds disturb wetting patterns.

Spray irrigation systems better utilise restricted growing areas, compared to surface irrigation, by not requiring productive land be used for supply channels, headlands, recirculation works, and so on.

Other features

Spray irrigation can be used, with adaptation, for frost protection in marginal frost conditions. The continuous application of water to the plant surface prevents the cell liquid freezing. (If misting or fogging is used, heat loss by radiation is reduced, also aiding in frost protection.)

Chemicals can be easily injected into a piped reticulation system, particularly fertilisers, but this can also be accomplished with other methods of irrigation.

Spray irrigation of effluent is feasible, as it is by other irrigation methods.

Spray irrigation equipment is better adapted to fire protection than other methods.

Types of spray systems

A wide range of different types of spray irrigation are in use.

Permanent installation

Permanent installation is expensive, since the equipment cost is dedicated to a fixed area and cannot be moved, but gives best control over irrigation management.

Permanent sub-mains and spraylines

It is common to permanently install underground mains and sub-mains to form the reticulation network of a spray irrigation system. It is less common to permanently install the sprayline except on high value perennial crops, playing fields, ornamental applications and greenhouses. Such systems are easily automated.

Movable sprinklers on fixed laterals

For turf and ornamental irrigation, it is possible to permanently install laterals, but use special bayonet type fittings to enable individual sprinklers to be moved from one station to another. This makes management simpler, and allows pipes to be permanently pressurised, but enables removal of sprinklers (for example, from playing fields) and better utilisation of sprinklers. (New installations would use "pop-up" sprinklers, which automatically retract into a buried housing when water pressure is closed off.)

Solid set system

This is a type of above-ground installation used in some annual crops where the spraylines and sub-mains are installed, without shifting, for the life of the crop. The equipment is installed after the last operation of ground preparation and sowing, then removed after the last irrigation before harvest. Special couplings and fittings are used, which are more easily assembled and dismantled in the field.

Portable systems

With portable systems, the spraylines or machines are moved between runs or stations. This enables the equipment to be used over a larger area, thereby reducing capital cost per hectare, but possibly increasing labour costs. The equipment is selected so that it completes its cycle of runs or stations before irrigation is required again at the start of the cycle.

Hand shift aluminium spraylines

Couplings at the end of each length of pipe enable quick attachment or removal so that the sprayline can be shifted manually to each sprayline position. Each coupling has a gasket which prevents leakage when the water is under pressure, and provision for attachment of the sprinkler or riser. Skids or stands are attached to the coupling to ensure the sprinkler is positioned vertically and/or to provide some height for the sprinkler to clear the crop.

Connection to the sub-main is accomplished by a short length of flexible pipe into a hydrant. The hydrant is just a

Figure 6.1 *Hand shift spray irrigation of pasture.*

stand-pipe and valve connected to the buried sub-main, with a coupling to suit. The flexible pipe can be long enough to enable two or three sprayline positions to be serviced from the one hydrant.

Hand shift systems incur a significant labour cost in shifting pipes, often under adverse paddock conditions. The systems described in the following sections are designed by various means to reduce labour requirements.

End tow

The sprayline is connected end-to-end by a steel cable, or by a special fixing method at each coupling, and each coupling is mounted on a skid. This enables the whole sprayline to be towed by the tractor. This is suitable only where the shifting can be accomplished in the same direction the sprayline runs, but can be useful in intensive cropping applications (such as vegetables).

Angle tow

This type of equipment is also sometimes referred to as "end tow". Each section of sprayline is supported by a pair of wheels which are able to pivot. The tractor pulls the sprayline from one end, in a 45° direction to the sprayline, to move it "crab-like" half way toward its next position. It is then pulled from the opposite end to complete the shift. This enables long spraylines to be quickly moved. This method is common in small to medium areas of pasture and lucerne irrigation. The sprayline will buckle if moved when full of water, so couplings contain a valve which opens to release water when the water pressure is turned off.

Figure 6.2 *End tow sprayline in asparagus. Note the skids beneath the couplings. Sprinklers should all be standing vertically, on risers sufficiently high that the water trajectory clears the crop height.*

Figure 6.3 *Angle tow sprayline on pasture.*

Figure 6.4 *Typical side-roll spray irrigation application.*

Side roll

The sprayline is mounted on relatively large wheels, such that the whole sprayline can be moved in a parallel direction to its next position. This movement can be done manually or the equipment can be fitted with an engine or water motor to provide slow continuous movement. Variable speed allows for alteration to the depth of water applied. Powered systems simplify moving the sprayline back to its start position at the completion of an irrigation cycle. Side roll machines can also be towed from the end to move the spray line to a different field, by the temporary use of transport wheels under each large rolling wheel. The diameter of the rolling wheel permits irrigation of taller crops, but traction over wet soil needs to be considered. With powered systems, each sprinkler is connected to a weighted connecting pipe, so that it always remains vertical as the sprayline rolls.

Travelling irrigator (soft hose)

This term refers to a family of irrigation machines, characterised by a single high pressure giant sprinkler (sometimes called a "rain gun") mounted on a carriage which is towed from one end of a run to the other by means of an anchored cable. A winch is located on the machine, which is turned slowly, by a turbine, piston or bellows type of water motor. As the winch turns, it winds up the cable, but because the cable is anchored at the far end, it has the effect of towing the machine along. Winch speed is adjustable to vary the rate of water application. Water is supplied to the sprinkler through a long flexible hose towed behind, but connected to a hydrant at the centre of the run.

The width of the run is governed by the trajectory of the water from the sprinkler, but can be up to 100 m. Adjacent runs are overlapped by around 30 per cent to give a somewhat smaller lane spacing. Length of run is determined by the length of hose that can be towed by the machine, but up to 400 m is possible.

Figure 6.5 *One type of travelling irrigator.*

Figure 6.6 *Typical spray pattern from a travelling irrigator, showing the substantial wetted width, and a good breakup of the spray accross the width.*

At the completion of the run, the sprinkler shuts down, with the cable fully wound up. To move to an adjacent run, the hose is wound onto a reel on the machine, power assisted by the tractor PTO on larger machines. The cable is re-anchored, the machine towed back to the hydrant, the hose connected, and the machine towed further back to the start position of the run.

These machines have the disadvantage of a high operating pressure, the possibility of structural damage to the surface of bare soil caused by impact of large droplets, the need for a special laneway to eliminate damage to tall crops by the dragging hose, and wind losses resulting from the long and high trajectory of the spray.

Hose-reel type travelling irrigators (hard hose)

In this type, the machine is kept stationary near the hydrant, and the rain gun is towed back toward the machine. This is

Figure 6.7 Top, *the reel of one hose reel machine and* bottom, *the sprinkler. Note the effect of insufficient water pressure on the distribution of water. A large proportion hits the ground at the end of the trajectory, without droplet breakup into a "curtain" effect.*

accomplished by winding up the pipe running to the sprinkler, rather than towing the machine by a winch and cable. Such machines are characterised by a large reel mounted on a carriage, which is rotated slowly by a water motor to wind in the sprinkler. Polythene pipe is used rather than lay-flat hose, and because of this, the machine is sometimes referred to as a hard-hose travelling irrigator.

From a crop point of view there is no difference compared to a conventional travelling irrigator, because the same type of rain gun is used. There are some possible management advantages, however:

- A dedicated laneway for the sprinkler carriage may not be required.
- The machine is easier and quicker to set up as it does not need an anchor cable.
- It is parked near the hydrant. It only needs to be moved from one hydrant to another at the end of each run and the sprinkler pulled to the start position away from the machine.
- Irrigation is accomplished on each side of the machine by simply rotating the reel 180° on its carriage.

Boom-type travelling irrigators

Some of the disadvantages of a conventional travelling irrigator can be reduced by replacing the giant sprinkler with a wide boom fitted with multiple sprinklers of lower operating pressure. This maintains a convenient width of run but reduces operating costs. Finer droplets and a more desirable wetting pattern are created. The boom is mounted on a similar type of carriage, propelled in a similar way. Moving the machine

Figure 6.8 *Low pressure boom type travelling irrigator*

between fields becomes more complex, requiring the boom to be partly rotated to be parallel to the run, and a person holding each end of the boom steady with a rope, during the transport operation.

Some machines are fitted with a rotating boom.

Centre-pivot irrigators

These self-propelled machines are designed for spray irrigation of large areas. They consist of a large diameter sprayline, supported on substantial towers, radiating from a central pivot point. Water is delivered to the pivot point in a buried main line, or the pivot point can be mounted directly over a bore. Electricity is delivered to the pivot point in underground cables, or is occasionally generated on-site.

Each tower is normally driven by its own electric motor, such that the machine as a whole travels in a circular fashion about the pivot, irrigating as it goes. The irrigated area is therefore circular, which will require adjustment to crop management practices. Travel speed is set at the outermost tower and is adjustable to allow for various application rates. The remaining towers are fitted with position sensors (of which there are various types) that instruct the towers to keep in a straight line. Should they not keep in line, because of bogging or malfunction, the machine is designed to shut down. This will happen also as a result of water supply failure.

The size and construction of the machine results in a sprayline high off the ground. Newer machines are fitted with drop tubes and low pressure rosette type sprayheads, that deliver water quite close to crop height.

Most centre pivot machines are permanently located in the field. However, mobile types are available that enable the

Figure 6.9 *Pivot end of a centre-pivot irrigator. Electricity and water are delivered to the pivot point, then along the machine. This older style machine has medium pressure sprinklers mounted on the sprayline.*

Figure 6.10 *A tower on a centre pivot irrigator. An electric motor, located in the can between the wheels, drives the tower forward when the position sensors, located at the top of the tower where sections of sprayline join, instruct it to move*

Figure 6.11 *Drop tubes on this machine bring water closer to crop height, minimising wind disturbance, and enabling selection of low pressure nozzles.*

Figure 6.12 *A corner watering extension is fitted to this centre-pivot machine. The last section of the sprayline can swing in and out to follow a non-circular boundary to the field.*

same machine to be used on multiple circles. Corner watering attachments are available (but not common because of the cost) to allow watering of the corners of a square field to maximise production per field.

Lateral move irrigators

The limitations of circular irrigated areas are overcome by the use of lateral move machines. These resemble centre pivot machines in construction and method of water application, but they move continuously in a straight line to water large rectangular areas. This type of machine can cover the largest area of all spray irrigation methods, with individual spraylines up to 1500 m long, travelling up to 4000 m, although smaller machines have been introduced to provide for more intensive production and/or overcome site difficulties.

With larger machines, the centre span of the sprayline carries an engine to drive a pump and an alternator. The

Figure 6.13 Top, *the centre of one lateral move irrigator. An engine driven pump draws water from a supply channel (its fuel tank is mounted on the other tower). The trailing tank contains fertiliser concentrate and an injection pump.* Bottom, *his view of the (empty) supply channel gives some impression of the area this machine can cover.*

alternator supplies electric power to the drive motors for each tower. The pump draws water from a channel, which runs the length of the centre line of the area to be irrigated, and delivers it to the sprayline which straddles the channel. Similar methods to centre pivot machines are used to keep the sprayline straight but additional, quite sophisticated, guidance systems are used to keep the whole machine tracking parallel to the channel.

The channel supplying water to the machine must be straight and would normally be constructed with only a slight gradient. This partly conflicts with the ability of the machine to negotiate gentle undulations, although some systems have been installed with stepped channels to accommodate steeper slopes.

Lateral move irrigators are available where water can be supplied to the sprayline from a large diameter hose, connected to hydrants along the edge of the irrigated area. Machines as

Figure 6.14 Top, *the drive carriage of a hose-pull lateral move irrigator and* botttom, *the trailing hoses.*

small as one span, with overhanging extensions, are available for more intensive production. These smaller machines can be towed from the end to additional sites, thereby increasing their application to fields of unusual shape and maintaining a low cost per hectare.

As with other spray irrigation machines, lateral move systems can be easily automated and give good control over water application.

Sprinkler design and performance

Sprinkler characteristics greatly influence sprinkler performance in achieving a satisfactory application of water.

Types of sprinklers

Sprinklers vary according to their method of generating droplets, and the water pressure at which they normally operate.

Medium pressure rotary knocker

Most agricultural sprinklers are of this type, designed with an operating pressure in the range 200–400 kPa (30–60 psi). Water is forced out of single or dual nozzles, breaking up into droplets as it leaves the nozzle. The water hits a rotating striker arm, which performs two functions:

- As the arm strikes the water stream, driven by a spring, it deflects some of the water to a smaller radius, improving the distribution of water near the sprinkler.
- The force with which the arm hits the sprinkler causes the sprinkler body to rotate part of a turn, thereby accomplishing rotation. Spring tension can be adjusted to vary the speed of rotation.

Various sized nozzles are available. Nozzle size(s) and water pressure at the nozzle can be selected to provide the required discharge rate and wetted diameter. This in turn determines the sprinkler spacing for optimum uniformity and application rate over the field. A sample sprinkler performance table is shown in Figure 6.15

The angle of trajectory of the sprinkler is normally from 20–25° from the horizontal. Special low trajectory types are available for watering under trees or in high wind sites.

Nozzle blockage can be a problem; for example insects and spiders can occupy the nozzle between waterings. Frequent checking for blocked nozzles is recommended, as it

Figure 6.15 *Typical sprinkler performance information.*

Performance Chart–Double Jet

Jet(mm)	5.6		6.0		6.4		6.8		7.2	
kPa	Dia (m)	L/m	Dia (m)	L/m	Dia (m)	L/m	Dia (m)	L/m	Dia (m)	L/m
200	29.4	22.7	29.9	26.5	30.3	28.5	30.7	34.9	30.9	38.0
250	30.5	25.4	30.8	29.2	31.2	31.4	31.6	37.5	31.8	42.0
300	31.6	28.0	31.8	32.0	32.1	34.5	32.5	41.3	32.8	46.5
350	32.3	30.5	32.7	35.0	33.0	37.8	33.4	44.8	33.6	50.5
400	32.8	32.2	33.6	37.5	33.8	40.1	34.2	47.7	34.5	53.8
Jet code	5164		5165		5166		5167		5168	

Performance Chart–Double Jet

Jet(mm)	5.6 x 3.2		6.0 x 3.2		6.4 x 3.2		6.8 x 3.2		6.8 x 4.8		7.2 x 4.8		7.2 x 5.6	
kPa	Dia (m)	L/m	Dia (m)	L/m	Dia (m)	L/m	Dia (m)	L/m	Dia (m)	L/m	Dia (m)	L/m	Dia (m)	L/m
200	29.5	32.2	28.9	36.0	29.2	38.0	30.2	43.4	30.2	53.8	31.2	58.0	31.4	64.5
250	31.4	37.4	29.7	40.0	30.4	42.0	31.3	48.5	31.3	60.0	32.0	65.0	32.2	72.1
300	32.2	40.7	30.6	43.7	31.8	46.5	32.0	53.3	32.0	66.0	32.6	71.2	33.0	79.0
350	33.5	43.7	31.5	47.4	32.7	50.4	32.9	57.5	32.9	71.5	33.5	77.2	33.8	85.4
400	34.2	46.6	32.4	50.5	33.6	53.8	33.8	61.2	33.8	76.7	34.5	82.2	34.8	91.2
Jet Code Main	5164		5165		5166		5167		5167		5168		5168	
Rear	5177		5177		5177		5177		5179		5179		5180	

Standard recommended jet is 5.6 x 3.2

Courtesy: James Hardie Irrigation Pty Ltd.

is for sprinklers failing to rotate or not standing in the vertical position. Sand and silt in the water may cause nozzle wear, but most high quality sprinklers are designed to resist damage to bearings and seals caused by the presence of sand. Plastic sprinkler parts may become brittle with time and therefore more susceptible to damage.

High pressure rain guns

A single high pressure (400–700 kPa) giant sprinkler is used on travelling irrigators and hose reel machines. It operates on a similar principle to a medium pressure type, but nozzle size and water pressure are much higher, giving a much larger wetted diameter and sprinkler discharge.

Low pressure sprinklers

Rotary knocker type sprinklers require a certain water pressure to drive the striker arm. Low pressure (100–200 kPa) sprinklers use other means to accomplish droplet generation. In many centre pivot and lateral move irrigators, water is directed onto a stationary "anvil", shaped to break the water stream into a fan type spray. Other devices include spinning discs and spinning arms, often used with microirrigation systems.

Others

Many horticultural and ornamental applications use a reaction type sprinkler. In this type, two nozzles pointing in opposite directions are carried on the ends of a relatively wide arm. Water leaving the nozzles causes the arm to rotate with the angle of trajectory of the nozzles variable to adjust the rate of rotation. Often, this same mechanism is used to provide power to a cable winch to propel the machine along.

Various types of pop-up and quick coupling sprinklers are available, mainly for turf and grounds watering.

Figure 6.16 *Sprinkler irrigation of roses. Many horticultural crops prefer not to have wetted foliage.*

Sprinkler performance criteria

These relate to the ability of the sprinklers to provide a uniform pattern of "rainfall" on the ground surface.

Droplet size distribution

A range of droplet sizes is produced as the water stream breaks up leaving the nozzle. Larger droplets will have a greater momentum and therefore travel further from the nozzle, but will increase the risk of soil and even plant damage (for example, to flowering parts). Small droplets may be preferred, but will be much more susceptible to wind drift and evaporation in the air, thereby reducing application efficiency.

Smaller droplets are achieved with higher pressures and/or smaller nozzle sizes.

Wetted diameter and sprinkler spacing

The diameter of the wetted area increases as nozzle size and pressure increase. The relationship is not linear, that is, a doubling of pressure does not result in a doubling of wetted diameter, since the effects of droplet size come in to play.

The wetting pattern from a single sprinkler usually results in a large variation in depth of application, from maximum at the sprinkler position to zero at the perimeter of the wetted area. The pattern is distorted further by the action of wind. Consequently adjacent sprinklers must be positioned so their wetting patterns overlap. For rectangular spacing of sprinklers, the distance between sprinklers should be 0.55–0.75 per cent of the wetted diameter. The same applies to adjacent runs of self-propelled machines. Measurements of the actual wetting pattern in the field gives more accurate recommendations.

Sprinkler discharge

The flow rate from the nozzle also increases as nozzle size and water pressure increase, with a marked variation possible. The sprinkler discharge for a given spacing determines the application rate to the field as follows:

$$\text{Application rate (mm / h)} = \frac{\text{sprinkler discharge (L / s)} \times 3600}{\text{spacing (m)} \times \text{shift (m)}}$$

where spacing is the distance between sprinklers along the sprayline, and shift is the distance between sprayline positions. In the case of self-propelled machines, the travel speed influences the application rate: it is easiest to refer to the manufacturer's specifications.

Relationship between the above

Sprinkler manufacturers publish performance charts that relate sprinkler discharge, nozzle size, water pressure and wetted diameter. Some sample data is included in Figure 6.15. As a starting point, it is necessary to decide the required application rate first (in mm/h). Sprinkler spacing is often governed by standard pipe lengths or standard models of self-propelled machines. This information can then be used to select the optimum nozzle size and water pressure to suit any given operational requirement.

Need for risers

Although the sprinkler can be attached directly to the sprayline coupling (see Figure 6.1), there are two reasons why the sprinkler should be mounted on a riser:

- to ensure the sprinkler is above crop height,
- to provide non-turbulent water flow into the sprinkler.

High sprinklers may become unstable and topple over. In permanent installations the riser may need to be staked. In portable systems the sprayline coupling could be fitted with a stabilising foot, or it could be fitted with legs.

Application rates

The maximum application rate should not exceed the infiltration rate of the soil surface. If it does, the water depth at the surface will increase and runoff will occur.

The application rate may be determined by the need to irrigate a given area in a particular time. For permanently installed equipment, there is usually no limitations. For mobile equipment, which must water a certain number of shifts or stations within the irrigation interval, the application rate must be sufficient to get the required amount of water on within the prescribed time. Given that a slow application rate has some advantages over excessive application rates, there must therefore be enough equipment installed, and of sufficient capacity to suit.

The advantages of slow application rate include:

- reduced runoff,
- maintenance of good soil surface structure,
- reduced labour per day (although total labour requirement for the irrigation is unaffected),
- more uniform infiltration,
- better match to the yield from bores.

Uniformity of distribution

The desired level of uniformity at the soil surface is difficult to determine, because of the effects of lateral water movement through the soil profile. Generally speaking, for a given sprinkler, nozzle size, pressure and wind pattern, there will be an optimum sprinkler spacing where the distribution of water between the sprinklers is most uniform. This spacing may then need to be modified to suit the equipment in use, standard pipe lengths, layout of the field and the pipe network.

Areas of the field which do not receive sufficient water will not reach their yield potential, and may ripen prematurely. Areas of the field which receive too much water will suffer yield loss due to temporary waterlogging. Consequently, there is an optimum depth of water to apply and it is desirable to minimise the variation from this.

The uniformity of distribution can be analysed by looking at sprinkler spray patterns, where a graph can be used to represent depth of water applied according to distance from the sprinkler.

If an ideal spray pattern was created, the spacing between sprinklers should be half of the wetted diameter; that is the next sprinkler should be located at the wetted perimeter of the first. This would result in wetting patterns overlapping, but the combined wetting pattern would be uniform.

However, in the field, irregularities and distortions occur, resulting in significant departure from an ideal wetting pattern.

It is also convenient to observe how depth of water application varies in both directions about the sprinkler, and depth charts can be constructed.

Figure 6.21 shows the position of a single sprinkler, and lines represent the contours of water depth. Simple measuring containers can be located in the field on a grid or radial

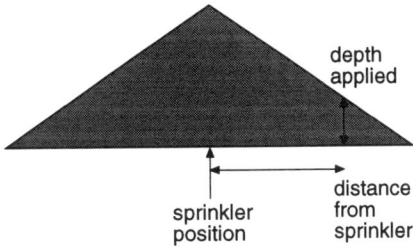

Figure 6.17 *Ideal spray pattern - single sprinkler*

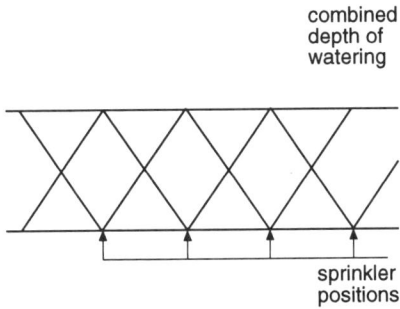

Figure 6.18 *Ideal spray pattern - Multiple Sprinklers, 50 per cent overlap*

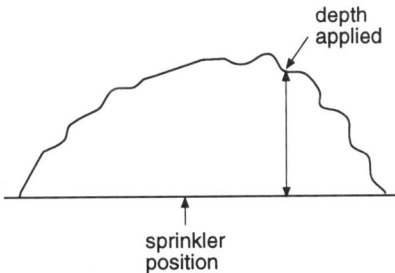

Figure 6.19 *Actual spray pattern - single sprinkler*

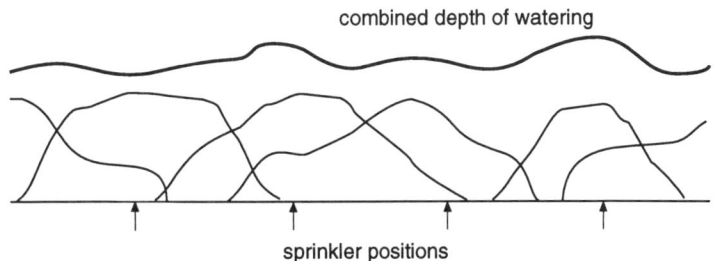

Figure 6.20 *Actual spray pattern - multiple sprinklers*

pattern under the sprinkler to measure such a distribution in the field.

Adjacent spray patterns can then be combined to represent the distribution between adjacent sprinklers, as an aid to determine optimum sprinkler spacing.

Sprinkler manufacturers sometime publish coefficients of uniformity for the wetting pattern of their sprinklers. Check that the tests are conducted by an independant organisation.

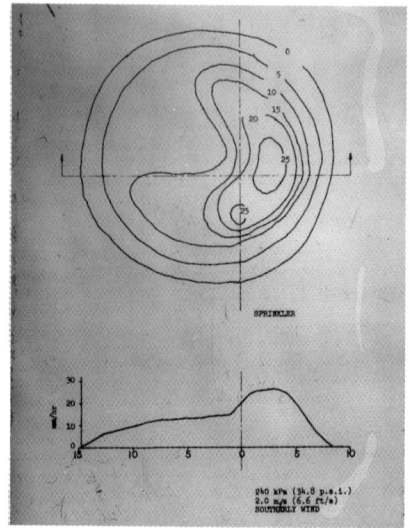

Figure 6.21 *Map of wetting pattern distorted by a wind speed of 2 m/s*

Summary of design procedures

1. The peak irrigation requirement and interval are determined, which in turn determines how much water must be applied and how quickly. This determines flow rate through the pipe network and pump output. Consideration should be given to crop and soil characteristics, and site topography.

2. Determine the discharge rate per sprinkler, based on irrigation amount and expected sprinkler spacing. In conjunction with this decision decide the nozzle size(s) and water pressure to provide the necessary discharge. Consult sprinkler performance charts to determine the optimum combination. For self-propelled machines, the run spacing and travel speed of the machine need to be ascertained, based on manufacturer's data.

3. Determine the diameter of the sprayline lateral, given the flow rate per sprinkler, the number of sprinklers, and the length of the sprayline. For self-propelled machines, this decision is pre-determined by the manufacturer, to suit the performance specifications of the machine.

4. Design the network of sub-mains and mains. Consider layout or positioning of spraylines or runs, distance from proposed pump site, fittings, and so on. Decisions need to be made on pipe material and diameter, method of installation, and optimum balance between cost and layout. Consider the effect that layout will have on pumping duty; for instance will the pump be able to handle the pressure and flow rate when the sprayline is furthest away from the pump, compared to its nearest position? This is particularly important to sites of irregular shape or layout.

5. Select pump and associated equipment. Consult manufacturers' performance data to help this decision.

6. Select power supply and associated equipment.

Maintenance

Apart from maintenance of the pumping station, which is discussed separately in Chapter 8, spray irrigation systems require only routine maintenance. Blocked nozzles should be cleared, and damaged sprinkler parts replaced. Spare parts should be kept on hand.

Damage could be caused by:

- Careless handling of the equipment; sprinklers, hydrants, above ground pipes.
- Wear caused by excessive abrasive material in the water.
- Normal ageing of seals and gaskets.
- Movement of buried pipes, and subsequent failure of points of connection in particular, usually caused by inadequate initial installation.

Self-propelled machines will require more substantial routine maintenance, particularly the lubrication requirements of gearboxes and bearings, but also attention to tyre pressures and electrical components. The owner's manual for the machine will have specific instructions.

CHAPTER 7

Microirrigation systems

Microirrigation refers to various systems of watering where attempts are made to supply water to more closely match the plant's daily requirements. Various types of emitters are used, located along polythene laterals, at a spacing matched to the soil type and plant spacing. The system is designed to operate for longer periods of time at low discharge rates. The emitter discharge rate is selected such that the plant's water requirement is replenished at frequent intervals. This constantly maintains the soil moisture content close to field capacity, minimising water stress. Emitters are designed to operate at low pressure, such as 100 kPa (15 psi).

Various types of emitter are available to suit the crop, irrigation schedule and soil characteristics. Most applications require permanent installation, but some systems are installed and removed each season (including drip irrigation of vegetables and cotton).

There are potential advantages of this type of watering:

- Increased yields are possible, since optimum soil moisture levels are maintained.
- Application efficiently is high (over 90 per cent), since the emitter is normally located near or at ground surface, minimising wind and evaporation losses. Laterals can be installed below ground, increasing efficiency even further. Application rates are sufficiently low and controlled that runoff and deep percolation can be eliminated. Consequently, limited water supplies can be utilised most effectively.
- Because most microirrigation systems are permanently installed, labour requirements can be low and the system can be easily automated (even by the use of soil moisture monitors to switch the system on and off).

- Plant foliage is kept dry, reducing fungal problems.
- Interrow spaces are kept dry during irrigation, reducing weed growth and creating minimal interference with mechanisation. Consequently, insect, weed and fungal control can be improved.
- Liquid chemicals can be applied through the irrigation water quite easily, with good control over dose rates.
- Microirrigation is suitable for steeper sites, where spray and surface methods are not feasible. Where long laterals are required, or where steeper slopes are involved, pressure-compensated emitters are available to control emitter discharge, and lateral diameters can be selected to give more precise control over water pressures along each lateral.
- In some applications, water which is too saline for spray and surface irrigation methods may be able to be used with microirrigation, as continuous water application can leach accumulated salts to outside the wetted root zone.

There are some limitations to the use of microirrigation:

- Cost is high.
- Most emitters rely on very small diameter orifices or flow paths to give control over discharge rates, and are therefore easily blocked. Filtration of the water supply is essential, and additional measures are necessary under certain circumstances. Further problems are that algae can grow in the system beyond the filter, clay colloids can be caused to flocculate, or that dissolved elements can pass through the filter and may precipitate due to various chemical processes. Consequently, water quality must be assessed thoroughly prior to installation, so that appropriate strategies can be considered. Emitter selection is very important.
- Poor watering techniques can be introduced. Overwatering is quite possible, even at very low application rates, particularly where sub-soil barriers inhibit water movement. Such problems are often not apparent at the soil surface.

Types of emitter

A range of emitter types is available to suit particular situations.

Drippers

Drippers are designed to give a slow application at frequent intervals, targeting the crop root zone. The drippers are located either on the outside of the lateral (on-line) or installed inside the lateral at manufacture (in-line). Laterals are installed permanently in the case of vineyards and orchards, either laid out on the ground or suspended from a wire (where they are more visible and clear of vegetation). In orchards and plantations of larger trees at wider spacings, multiple drippers are used around each tree, attached to an offtake from the lateral. In the case of row crops, the laterals are laid out for the duration of the irrigation season, then wound up on a large reel for later re-use. For plants grown in containers, drippers can be attached to the lateral by a short length of microtube. It is also possible to install in-line dripper laterals below the ground, which may have application in landscape irrigation, lucerne and other crops, but the risks of root penetration and the lack of visibility of the dripper may deter some users.

For standard drippers, there is a particular relationship between water pressure in the lateral and dripper output. To achieve a uniform output for each dripper in the field (a flow variation of 10 per cent or less is usually acceptable), it is necessary to provide a reasonably uniform water pressure along and between each lateral. Consequently, a drip irrigation system requires careful hydraulic analysis of the pipe network to ensure uniformity. This is difficult to achieve for long laterals and steeper sites, but pressure compensated drippers are available. These are manufactured with a small diaphragm that automatically adjusts the dripper orifice size to provide a constant output over a range of water pressures. A high degree of manufacturing accuracy is necessary to ensure this occurs.

Figure 7.1 *Drip irrigation of winegrapes. The dripline is carried on a wire above the ground. The wetted surface area is visible.*

Figure 7.2 *In this nursery, each container receives water from the lateral through a small peice of microtube.*

Figure 7.3 *Drip irrigation of corn. The driplines, connected to hand-shift aluminium sub-mains, are wound up each season and re-used.*

Dripper spacing needs to be matched to the soil's hydraulic conductivity and dripper output, and the peak irrigation requirement of the crop must be met within the required irrigation interval. In-line drippers can be selected to suit a wide range of standard spacings. On-line drippers can be installed at any position in the lateral, provided the water pressure along the lateral is kept reasonably uniform.

Blockage of drippers is a major risk, because of the small orifice size used. Manufacturers go to great lengths to minimise the risk of blockage. Rather than a single pin-hole type of orifice, the flow path through the dripper should be a tortuous one, because this enables a short but larger diameter flow path. Also, the flow through it is more likely to be turbulent, (rather than laminar) which will help keep suspended material moving through the dripper. The inlet to the flow path should have its own screen to prevent particles entering. Some on-line drippers can be disassembled and cleaned manually.

Other factors to consider in dripper selection include:

- The ability of the dripper (and diaphragm) materials to resist chemicals in the irrigation water.
- The uniformity of output and range of operating pressures for pressure compensated drippers.
- Temperature changes on dripper performance (as well as expansion and contraction of laterals).

Perforated or permeable tube

There are various versions of perforated tube available where water is delivered through a pin hole through the wall of the tube, with perforations made at fixed intervals along the length of the tube. The holes are relatively small and easily blocked, so some types utilise secondary chambers built into the tube to allow a larger hole to be used.

Permeable tubes are also available, where water oozes continuously along the full surface of the tube. One type is manufactured from a plasticised paper material, and another from a special re-cycled rubber which gives to release particles which might cause a blockage.

These may have application in vegetable, flower or greenhouse crops and can be buried below the soil or growing medium.

Microtubes

These are pieces of small diameter plastic spaghetti tubing inserted into the wall of a lateral. Discharge rate is governed by the length of microtube and occasionally a spreading device is inserted on the end. Their relatively narrow diameter can result in easy blockage, and although coiled types are available, the microtube can get in the way and be damaged or pulled out. They have been mostly superseded by other emitters.

Microjets

This term describes a family of emitters which spread water over a much larger surface area than a single dripper. This may be necessary in larger tree crops or ornamental applications, or where a higher discharge rate is required. Microjets are in common use in domestic garden watering.

Various types of fan type spray patterns are available (full circle, part circle, directed jets, mist). Most use small perforations in the top of the jet to produce the wetting pattern, or alternatively, the water stream strikes a stationary anvil to cause spreading.

Microjets have no moving parts (which distinguishes them from minisprinklers). Operating pressure is nominated by the

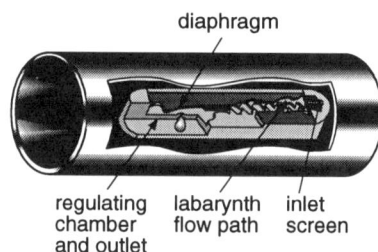

diaphragm

regulating chamber and outlet   labarynth flow path   inlet screen

Figure 7.4 *Cross section through one type of in-line dripper, indicating the tortuous flow path designed to help prevent blockage.*

Figure 7.5 *Microjet irrigation*

manufacturer and discharge rate is selected by choice of orifice size in the base of the jet. Discharge rates up to 200 L/hr are typical, with wetted diameters to around 2 m. The wetting pattern at the surface is often uneven, but this may not be a problem with low application rates, considering lateral infiltration in the root zone. Microjets need to be clear of vegetation often encountered at ground level, and need to be oriented vertically. Although they can be inserted directly into the wall of a lateral, they are best installed on a short riser or stake, with a connecting tube from the lateral to the jet.

Blockage of microjets is still a potential problem, since water must pass through a single, small-diameter orifice.

Minisprinklers

This describes another family of emitters, which use a rotating component to spread water over a larger area (2–5 m). This type of emitter also delivers a higher rate of discharge. Orifice size is often larger, so blockages may be less likely.

Many different types are available. Some use a spinning disc or cap, others a rotating arm or arms, and others resemble small versions of conventional sprinklers. Selection is based on crop requirements for frequency and amount of watering, operating pressure and discharge rate, wetting pattern, and emitter spacing.

Water quality and filtration

Filtration is essential with microirrigation systems, to remove particles suspended in the water. Unfortunately, filters alone cannot cope with all possible water quality problems, so

Figure 7.6 *Minisprinklers have moving parts to spread water over a larger wetted area whilst maintaining reasonably low operating pressure.*

additional treatments may be necessary. The choice of filter will depend on:

- The nature of the suspended particles and their concentration.
- The flow rate and pressure that must be accommodated.
- The diameter of the emitter orifice or flow path.
- The ease with which the filter must be cleaned, and the amount of water required to do this.

Types of filters available

Because of the importance of filtration, the different specifications available, and the range of possible problems with water quality, expert advice should be sought.

Wire mesh or gauze

These need to be of a large surface area and relatively fine mesh size to suit microirrigation applications. They are generally used as back-up filters, although self-cleaning versions are available. The mesh provides only a single surface to remove particles, and particles of irregular shape or those that are compressible might still pass through. Cleaning can be difficult, because the particles may be lodged hard into the mesh.

Centrifugal or spiral filters

These do not have a screen of any type, but cause the water to spin in passing from inlet to outlet. Particles are removed from the water by centrifugal force, collecting at a disposal outlet at the base of the filter. These can be used as initial or primary filters to reduce the load on the main filters, particularly when sandy material is present in the water.

Disc type

These consist of many plastic discs pressed together in a cylinder. The surface of the discs is manufactured with angled grooves at opposing angles, so that water passes through a tortuous pathway of intersecting grooves between adjacent discs, removing particles as it travels. This provides a longer pathway for filtration to occur, compared to a screen, with different disc and groove patterns corresponding to a range of equivalent mesh sizes.

Multiple cylinders are connected together to give the required flow rate. Cleaning is accomplished by loosening the discs and flushing with clean water. Self-cleaning types are available, which use only a moderate amount of water for backflushing.

Figure 7.7 *Multiple disc-type filters, mounted in a tower. This version is self-cleaning.*

Media filter

Water is passed vertically through a large tank containing graded gravel or sand. Multiple tanks are used to provide the required flow rate. Although the most expensive, these are generally considered the most effective type, particularly where the water supply is stored in dams. Although routinely specified for microirrigation systems in Australia, some irrigators have experienced problems with media filters being clogged very quickly, requiring frequent backflushing, which could be a result of improper selection of filter size and capacity. Some systems could benefit from some form of pre-cleaning equipment. In some cases, the filter could be removing material which will pass through the dripper anyhow, which could overwork the filter.

Figure 7.8 *Media filtration unit. Multiple tanks are necessary to accommodate the required flow rate, and to provide for automatic backflushing.*

Cleaning is achieved by temporarily reversing the flow through the tank (backflushing). During this operation, the media is suspended, releasing contaminants. A reasonably large amount (and velocity) of backflush water is required, normally going to waste. Automatic backflushing is available; as the filter commences to collect suspended material, there is an increase in the pressure required to push the water through. This results in a larger difference in pressure between upstream and downstream sides of the filter, which increases over time. When the pressure difference reaches a pre-set limit, the filter is automatically switched to backflush mode. Whilst being flushed, the filter cannot be in normal working mode, so installation requires at least two modules.

A back-up filter is required immediately after the media tanks to capture any media that might escape the tank.

Aspects of water quality

In the event of a pipe repair in the field, there is a high risk of material entering the pipe network after the filter. Flushing will be essential immediately after repair, and some projects have manual clean back-up filters installed at the inlet to each block.

Unfortunately, filters are only best at removing suspended particles, and may not remove other problem material that may be present:

- Algae present in the water supply may accumulate and grow in or downstream from the filter. Periodic chlorine treatment may be necessary, with the concentration of chlorine selected to suit the particular situation. Sometimes, continuous chlorine injection may be required, for example where wastewater is used.

Figure 7.9 *In this system, a manual clean disc filter is installed with the valve at the entrance to each irrigation block as a back-up to the main filters (system under construction)*

- Clay colloids are small enough to pass through the emitter, but certain chemical reactions can occur with other elements dissolved in the water, which causes the clay particles to flocculate. Some chemicals can be flocculated in storage reservoirs prior to use. Salts of iron, common from bore water, will react with oxygen, hence it may be necessary to oxygenate the water prior to use.
- Bacteria can form a slime inside pipes, which can detach from the pipe wall and cause blockages. Chlorine treatment may be necessary.
- Acid may need to be injected to lower pH and to control build up of calcium based deposits.
- Temperature fluctuations can initiate chemical reactions, although burying the pipeline will minimise hot temperatures. Fertiliser injection should be done before the end of the irrigation to enable clean water to flush the chemical through.
- Insects crawling into the emitter can be a problem. Keeping the emitter above ground level helps, as does periodic flushing with lateral ends open. Some emitters are designed with "anti-ant" closing devices.

Flushing of the laterals should be a regular practice; at the commencement of each season, and during the season as required. For manual flushing, the lateral end is folded onto itself and held by a clip. The clip is removed for flushing, then replaced. Laterals can be fitted with automatic flushing valves that release some water on every start-up. For larger projects, where flushing of individual laterals is inconvenient or too expensive, blocks of laterals can be connected into a common flushing sub-main, the whole block flushed by opening one or two valves. A reasonable water velocity is required during flushing, to dislodge and move the contaminating particles. Sub-main ends should also be fitted with a flushing valve.

Material may also accumulate in mainlines, either a gradual build-up of fine material, or through a pipe repair. Each irrigation block can be fitted with a back-up filter, but the low points of the mainline should be fitted with a scour valve to enable flushing of the mainline.

Fertigation and chemical injection

"Fertigation" is the injection of fertilisers into the irrigation water. Although this is possible with other methods of irrigation, it is a common and easily achieved practice with microirrigation systems. Being applied with irrigation, the fertiliser is easy to apply, with little extra labour and no extra

Figure 7.10 *One type of fertiliser injection unit*

machinery. Provided the hydraulic design is adequate, the fertiliser is applied uniformly to each plant, targeted at the root zone. Fertiliser can be applied at frequent intervals and at any required concentration, to suit plant requirements.

Nitrogen, phosphorous and potassium fertilisers are available for use by fertigation. Decisions need to be made regarding the type of fertiliser to use, the quantity to be applied, the effects it might have on equipment and soil, and its compatibility with other dissolved or suspended materials in the irrigation water. Soil and water testing should be thorough to minimise potential problems. Equipment and pipelines should be flushed with clear water at the end of fertilising, before the end of the irrigation cycle.

Fertiliser concentrate is mixed in a separate storage tank. Thorough mixing is required. For small projects, this may be just a large drum, with initial mixing by hand. Larger projects, where larger quantities and volumes are involved, will require larger storage tanks, and preferably continuous agitation by recirculation.

Various types of injection equipment are used to deliver fertiliser concentrate into the irrigation mainline. The most common use:

- Differential pressure across a valve to divert some water into a pressurised concentrate tank, then deliver it into the mainline.
- Differential pressure across a valve to divert water into a venturi suction device, which then draws concentrate from the storage tank before delivery to the mainline.
- Positive displacement (diaphragm or piston) injection pump with electric or hydraulic drive. Sometimes the speed can be varied.

The latter is the most expensive, but gives greatest control over fertiliser application and therefore more accurate and flexible management. Other factors in the choice of system include suitability to automation, effect on pressure in the mainline, resistance to corrosion, maintenance requirements, and whether fertiliser is to be injected in proportion to water flow through the main on a continuous basis, or in separate, defined doses (so that injection rate is independent of irrigation rate).

Chlorine may need to be added to the system, to control algae and bacteria, and acid injection may be needed to control calcium precipitates. Because of the risks associated with handling such chemicals, expert advice should be sought to determine the best form of chemical, the required dose rate and regime (continuous, intermittent or slug), and the equipment and technique to do the job. The need for this should be considered in the initial design.

Thorough analysis of water quality will help predict the need for these treatments. Apply treatment as a preventative measure before problems escalate and risk emitter blockage. Treatment can restore blocked emitters, but this requires high concentrations of chemicals to remain in the pipeline to enable cleaning to occur. Thorough flushing must follow.

Automation

The cost and level of management required for a microirrigation system warrant the use of an irrigation control system. Even for systems of only a few blocks, relatively inexpensive controllers add great flexibility and automation capacity. Highly sophisticated controllers are available for larger installations which can also log and analyse data collected during irrigation. Reports from such systems act as a monitor over system performance.

Laterals are grouped into blocks, each supplied by a submain. The block size is determined as much by flow rate and uniformity criteria as by planting area and soil type, although it is convenient if they are the same. Flow into each block is controlled by a solenoid valve, operated by low voltage electricity. A controller sends signals to the valve to open and close according to a program keyed into the controller by the irrigation manager. The programme would normally nominate the time and duration the valve opens, and the sequence required to irrigate multiple blocks. The program can be integrated with calculations of evapotranspiration from weather station data.

Fill cap with silicon and snap into position to ensure a waterproof connection

STATION 1   STATION 2   STATION 3

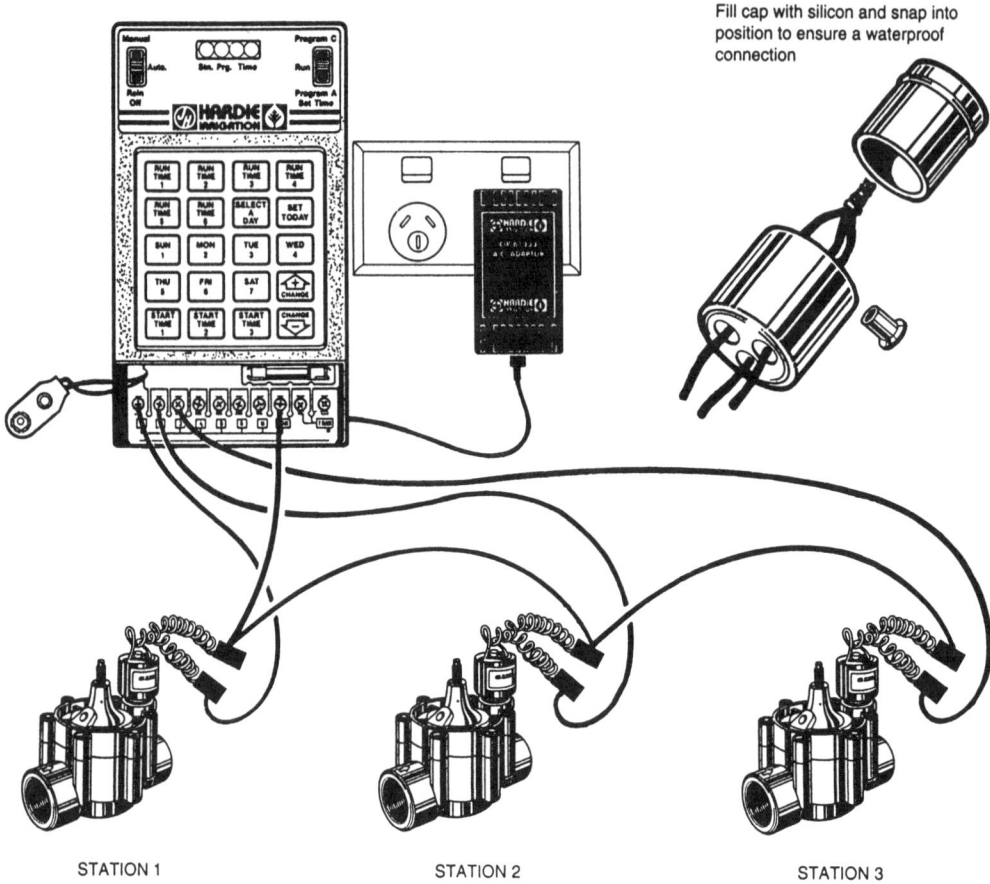

Figure 7.11 *The simplest form of controller instructs the valves to open and close to a programmed schedule.*

Signals are usually transmitted to the solenoid by wire cable(s) buried with the pipelines, although radio controlled equipment is available. Lightning and wiring defects remain the biggest technical problems. Some systems use multiple wires (one to each valve plus one common to all of them) and others use a communication cable of only two wires for all the valves.

Summary of design procedures

Microirrigation is typically applied to intensively managed high value enterprises, so a high degree of planning is justified. Also the system is able to accurately control water applications to individual zones or parts within the field. A detailed soil assessment is conducted at the planning stage, by the use of a large number of soil pits throughout the proposed area. The readily available water at each pit is determined along with other soil characteristics. This will allow accurate matching of plant growth characteristics and water requirements to soil variation throughout the area, and the detection of sub-soil problems. Climatic information is assessed to determine the peak irrigation requirement.

A detailed contour map of the site is constructed, including location and elevation of pump stations, and preferred planting sites noted.

The optimum layout of the irrigation system can then be drafted, largely based on the size and layout of crop units that need to be watered as an individual block. If the blocks are large, they may need to be divided into more than one irrigation station so that uniform water application can be achieved. The location of crop rows dictates the alignment of laterals, but the supply to them from sub-mains and mainlines can be varied to suit hydraulic and economic criteria.

Pipe diameters are selected based on the peak flow rate to the various irrigation stations, the allowable flow or pressure variation permitted within each block, and costs. Irrigation designers use computer software to determine the optimum design. Pump specifications are then determined. Fertiliser injection, chemical injection, filtration, controller, flow meters, air valves and scour valves are added to suit the situation.

Maintenance

A high level of maintenance is required for microirrigation systems. The need to keep laterals and emitters free from blockage and the need for filtration has already been discussed, as has chemical treatment and flushing. This may be required on a regular basis throughout the irrigation season to prevent major problems occurring.

A complete pre-season check is necessary. The operation of mechanical, electrical and hydraulic equipment should be checked, and the system flushed and pressurised to check for leaks. Performance and condition of filters and check filters should be checked.

During the season, frequent pressure checks should be made in the field, to ensure water output is as it should be. Sub-main valve stations should be equipped to enable use of a portable pressure gauge to check pressures on the spot. Check dripper output occasionally by measuring the volume over a one hour period, and compare to specifications. During routine operations, check for leaks, noisy valves and pumps, and the operation of drippers. For example, if the dripper starts to squirt instead of drip, it is possible that partial blockage could be commencing. A noisy pump could be a mechanical problem, but it could also be cavitation, a condition created by excessive suction lift or restriction to water flow entering the pump.

As part of a preventative maintenance schedule, inspect filters and flush the system on a regular basis.

Most major components (filters, pumps, valves and so on) will require an annual service at the end of the season. Details should be supplied by the equipment manufacturer.

CHAPTER 8

Reticulation

This chapter provides an introduction to the design and installation of pump stations, pipelines and channels used to reticulate irrigation water from storage to point of application. These are specialised topics requiring expert advice, so this chapter only provides the basic principles upon which detailed design is based.*

Delivering the water to the irrigated area can be accommodated by the use of open channels (where water flows by gravity) or closed pipelines (where a pump pressurises the water). Channels are typically associated with gravity irrigation methods, but some spray and microirrigation systems use a pump to draw water from a channel. Pipes are typically used with pressurised irrigation methods, but are occasionally used to deliver water to surface irrigation systems where water losses from channels are excessive, or land area is limited.

Pump stations

Site selection

The following criteria should be considered:

- Proximity to irrigated area; pumps should be located as close to the irrigated area as possible, to minimise the distance and therefore the cost of delivering water to the crop. The pressure required at the pump will be determined by both distance and elevation between water source and point of application, which will partly determine power requirements.

---

* The *Farm Water Supplies* book in this series provides some additional information on pipeline design and selection of pumping equipment.

Figure 8.1 *On this vineyard, the water supply and pump station are conveniently adjacent to the irrigated area, minimising costs of water reticulation. Water is delivered first to the top of the hill, then submains bring the water downhill to the laterals, to ensure even pressure to the laterals.*

- Proximity to water source; the pump(s) need to be as close to the water as possible. There is a distinct limit to the ability of a pump to create suction, so the water level can only be a maximum of 5–8 m below the pump, depending on circumstances described in more detail on page 124 (less is preferred). Axial flow pumps cannot create suction at all, so the pump impeller must be submerged below water level by a certain distance.

  This creates installation problems where the water supply has a fluctuating water level and with large dams, where the depth of useable water exceeds about 6 m. It may be necessary to consider mounting the pump on rails so it can be lowered closer to a receding water level, using a submersible type pump (either floating or mounted on a rail) or to install an outlet pipe at the base of the dam wall. This latter approach requires great care during installation to prevent possible leakage of water along the outside of the pipeline.

  In the case of river installation, it is necessary to consider protecting the installation against a rapid increase in water level during flood events. Again, a number of options are available. The most common are to shaft drive the pump from an engine or motor located well above flood height, or mount the equipment on rails, and raise it above flood height before it arrives. The latter approach requires a timely decision.

- Proximity to electricity; for most installations, electricity would be the preferred power supply (compared to diesel

Figure 8.2 Top, *riverbank axial flow pump, electric motor housed above flood height, shaft drive to the pump;* bottom, *water is delivered to the highest end of this open channel*

engines), but the cost of bringing electricity to the pump site can be quite expensive. For many pump sites, diesel power is the only option.

- Physical risks to the pump station; the need to protect engines and motors against flood inundation has already been noted (the pump itself can tolerate being wet). Other risks include physical damage from debri floating down rivers, particularly during floods, and erosion, scouring or settlement of the foundations under the pump station, caused by inundation during floods, or accidental events such as a pipe bursting near the pumphouse.

Pump duty

The first step in pump selection is to determine the duty the pump must perform. This is determined by the flow rate the pump must deliver, and the pressure against which it is delivered.

The flow rate is determined by the application rate to the blocks, summed according to the size and number of blocks being irrigated at any one time. A high flow rate will require a larger pump and more power, but offers the greatest potential to meet peak water demand.

Where the flow rate is variable, multiple pumps may be the best option, switched in and out according to demand.

The pressure at the pump is commonly referred to by the term 'head'. The head is measured in metres. (Pressure is measured in kilopascals or kPa, and kPa ÷ 9.8 = metres head.) Head is equivalent to the vertical height a column of water could be held at that pressure.

The head on the pump is influenced by three factors, the vertical distance between water source and the highest point of the delivery system (the static head), the pressure required at the point of application (for example at sprinkler nozzles or microirrigation emitters; the pressure head), and the pressure required to overcome the resistance to flow caused by friction in pipelines (the head loss from friction). Each of these components is considered separately for each situation, and summed to give the total head. For a given flow rate, a high head requires more power.

The irrigation designer only has partial control over these factors. Flow rate is largely dictated by the peak water requirement of the crop, and the minimum interval between irrigations. The static head is dictated by the relative locations

Figure 8.3 *Multiple pumps are commonly used to meet a variable demand. These are centrifugal pumps, direct coupled to electric motors selected for power output and speed of rotation to suit the pumps. Note the flexible couplings between pipe and pump flange, to minimise transfer of vibration and misalignment to the pipe network in this pump station. The pipe manifold has provision for an extra pump to be added. (Courtesy: Central Highlands Management)*

of irrigation blocks and water source. The pressure head can be varied by choosing a low pressure method of irrigation, compared to a high pressure one, but once the irrigation method has been chosen, the pressure at the nozzle or emitter is selected to give the required output. However, there is some control over head losses from friction, and it is desirable that there be kept reasonably low. This is done by choosing a sufficiently large pipe diameter that friction is acceptably low at the design flow rate.

Pump type

These can be classified into two main groups, based on the way in which mechanical energy is transferred from the pump to the water.

1. Hydrodynamic pumps, such as:
   - Centrifugal (or radial flow)
   - Axial flow
   - Mixed flow.

   These pumps are characterised by the transfer of energy to the water from a rapidly rotating impeller. They can operate for short periods with the discharge closed, but generally require priming prior to start-up (that is the pump and the suction line must be filled with water). They are also characterised by the fact that when the pump pressure is changed, so is their discharge. They are in common use for irrigation applications.

2. Positive displacement pumps, such as helical rotor pumps. With positive displacement pumps, the same quantity of fluid is delivered for each revolution of the pump shaft, regardless of the pressure. Because of this characteristic, some form of pressure relief device is required.

Other classification systems are sometimes used. This one attempts to relate pump performance to practical considerations.

Each of these pump types is best suited to a particular duty. Positive displacement pumps frequently have a pulsating discharge, and so are generally unsuitable for irrigation applications. The exception is the helical rotor type, which has a continuous discharge, and may be suitable for high pressure-low flow duties (for example, reticulating water to a high pressure travelling irrigator).

Centrifugal (or radial flow) pumps are widely used. They are of low cost, with few moving and wearing parts, keeping maintenance costs down. They can be selected to suit a wide range of duties and applications, and have a constant, non-pulsating discharge. Needing to operate at high rpm, they are suited to direct coupling to electric motors, but are equally suited to belt drive applications. They are generally quiet in operation, and of compact size, and the pump can be aligned in a variety of directions.

Disadvantages include the need for priming and the lower efficiency compared to some other types. Suction conditions are critical to performance, and may be a limitation to their selection.

The impeller of an axial flow pump resembles a ship's propeller, and is mounted near the end of a tube, submerged under water. As the impeller turns, driven by a long shaft from an electric motor or diesel engine, water is pushed up the tube. The entry of water through the intake is guided by a set of stationary vanes to minimise turbulence and vortex formation that might damage the impeller. Stationary guide vanes are also located immediately after the impeller, again to minimise turbulence and to straighten the flow parallel to the delivery pipe.

The axial flow pump is capable of pumping extremely large volumes of water, but can only develop very small pressures.

There is a whole range of types that are neither true axial flow pumps nor true radial flow pumps. These are known as mixed flow pumps because their flow characteristics, and therefore their performance characteristics, are a mixture of axial and radial flow types.

Mixed flow pumps are sometimes used where the lift requirement exceeds that of an axial flow pump, but high flow rates are still required.

Pump efficiency and power requirement

For any combination of discharge and total head, there will still be many makes and models to choose from. Some can be eliminated because of specific site conditions (such as suction lift too high, borehole applications, risk of silt or abrasive particles in the water). Of those remaining, the ones which meet the required duty at a high efficiency will be well matched to the application.

Unfortunately, the efficiency varies according to the exact operating conditions of the pump and can be a little elusive to determine. If the duty is known, the efficiency can be obtained from the manufacturer, or from the performance data sheet for the particular pump. This will also nominate the required speed of rotation (rpm) of the pump shaft.

The amount of power required to run a pump under a particular head and discharge is given by the following formula:

$$Ps = \frac{Ht \times Q \times 9.8}{Ep \times 100}$$

Where: Ps = shaft power required in kW
Ht = total head the pump is working against in m
Q = discharge in L/sec
Ep = pump efficiency.

(Note that Ht × 9.8 = pump pressure in kPa)

Because the shaft power is dependant on both head and discharge, a change in either of these will affect the power requirement. For example, the total head often varies as water levels rise and fall, or as irrigation spraylines are moved across a field. For the calculation of power requirements, the maximum head condition must be met, since there will be sufficient power for all other conditions. However, pump efficiency also varies as head and discharge vary, and it is wise to check this for both maximum and minimum head conditions.

The shaft power of a pump is its input power, that is the amount of power that needs to be delivered into the pump shaft, by the motor or engine driving it. Because of inefficiencies in the method of transmitting power to the pump, and of certain factors pertaining to the engine or motor itself, the power rating of the engine or motor must be higher than the pump shaft power.

Figure 8.4 *This is a portable (skid-mounted) diesel powered centrifugal pump delivering water from a channel to a hand shift spray system. Although it looks a bit untidy, it is fitted with electric start and shut-down protection. It should be placed on a firm level pad*

In the case of direct coupled electric motors (provided the supply of electricity is adequate), the motor operates at high efficiency, and 10 per cent is added to the calculated shaft power to give the motor size.

In the case of internal combustion engines, many more factors (known as de-rating factors) need to be considered; the engine manufacturer's rated power must be adjusted, according to the following guidelines:

- Altitude and temperature; the manufacturer tests engine performance under standard conditions. Both ambient air temperature and altitude affect the density of air, which determines the amount of air that can be drawn into the engine's combustion chamber (assuming the engine is normally aspirated, not turbocharged), and hence the power developed. The loss of potential power is substantial at high temperatures and altitudes. Allow one per cent for each 3°C above 20°C, and one per cent for each 100 m above sea level.

- Cooling method; a water-cooled engine may use up to 5 per cent of its rated power in driving its water pump and fan, compared to an air cooled engine. This factor is difficult to estimate accurately, and often disputed since some air cooled engines have large fans, too.

- Continuous service; an engine is able to develop maximum power for short periods, but continuous operation at maximum power could result in accelerated wear. Continuous power output is generally considered to be up to 15 per cent less than rated power output, with this allowance is reduced in proportion to the hours per day in continuous usage.

- Age; as engines get old, performance drops because of normal wear of rings, bearings and so on. In addition, engine tune plays a large part in performance. An allowance of 25 per cent is usually considered necessary to account for these losses over time.

The required speed of rotation of the pump must also be accommodated by the power transmission components. Pumps can be directly coupled or belt-driven by either electric motors or internal combustion engines. If directly coupled to an electric motor then the pump can operate only at electric motor speed (2900 and 1450 rpm being the two common speeds), but if direct coupled to an engine, the pump speed will be the engine speed up to the rated speed of the engine. It is important that the pump rpm is that specified by the manufacturer. Belt drives enable greater variations in the choice of pump speed, via the selection of appropriate pulley sizes, but result in a small loss of power.

The following suggestions are a guide to the amount of power lost with various methods of power transmission:

direct coupled:	up to 2 per cent
shaft drives:	up to 3 per cent
V-belts:	up to 5 per cent
flat belts:	up to 10 per cent

Pump installation

Pumping equipment needs to be installed on stable foundations to prevent movement and minimise vibration. This is usually accommodated by mounting the pump and its drive unit on a rigid steel frame, and securing the frame to concrete foundations.

The steel frame should incorporate some means to adjust the alignment of the pump shaft. Pumps can be selected to operate with horizontal, vertical or inclined shaft, but the shaft bearings must be designed to suit, and the installed alignment should accurately match the manufacturer's recommendations. In addition, the alignment of the drive shaft, belts or coupling between pump and power unit will also require adjustment. Ensure the drive to the pump is correctly guarded to ensure operator safety.

Shaft alignment should be double-checked prior to commissioning of the pump station, and frequently during initial operation. Alignment can change as a result of vibration, and strains due to temperature effects or pipeline movements. A range of shaft and pipe couplings are available to absorb such effects, and accommodate small amounts of misalignment. Pipelines should be supported independently of the pump.

Protection should be provided to the engine or motor driving the pump. Engines should be fitted with a "watchdog" to shut the engine down in the event of engine malfunction (due to overheating or loss of lubricating oil pressure) and malfunction of the irrigation systems (loss of water pressure). Electric motors are fitted with thermal overload protection as standard, but they can be fitted with quite sophisticated control systems to monitor system performance and respond to a variety of sources of malfunction as required.

The starting method of large horsepower pumps needs to be considered. The current drawn by a large electric motor during starting greatly exceeds that required for normal running. Electricity supply authorities will normally require the installation of a reduced current starter or starting procedure.

A sudden pump start can also create undesirable pressure surges in pipelines, as can sudden pump shut-down, and rapid valve closures. Pressurised irrigation systems should be designed with these risks in mind, referred to as water hammer. Pipe diameters should be selected to keep water velocity to an acceptably low level, and automatic valve and pump operation should be "soft" (that is the water velocity changes slowly). Manual valve operations should be performed slowly. Pressure surges can also cause sudden movement or failure of pipelines.

The suction side of the pump plays a critical part in the overall performance of the pump. Performance drops markedly if there is a fault on the suction side. A pump can lose its prime if there is a leak in the suction pipe at a bend or join (this will result in air leaking into the suction pipe, not water leaking out, since the pressure inside is less than outside). Always install the suction pipe so that it does not have any high points in it (so that it rises continuously into the pump) to prevent air bubbles accumulating and the pump losing prime. Minimise turbulence in the suction pipe and pump inlet by using large diameter fittings and large radius bends.

Avoid blockages at the inlet. Put the inlet on a float to keep it out of the mud, and provide a screen to prevent debris accumulating. If the water cannot get into the pump, or if the suction lift is too high (greater than about 5 m for centrifugal pumps), then cavitation may occur.

Cavitation is a phenomenon which results eventually in damage to the pump or pipeline and also markedly reduces pump performance. Although other pumps can be subject to cavitation, centrifugal pumps are particularly susceptible. Cavitation may occur if the supply of water into a pump is severely restricted whilst it is operating. If the suction pressure inside the pump exceeds the theoretical maximum of one atmosphere (10.35 m) the water starts to form tiny gas bubbles. In practice it can occur at suction heads closer to 8 m when dissolved gases come out of solution. Because of their susceptibility, the suction head for centrifugal pumps should be limited to around 5 m. Cavitation causes a reduction in performance, and often a loud noise and vibration, which is a good indicator of a fault. Further, as the gas bubbles approach the high pressure side of the pump, they are forced to implode. This can occur with surprising force, sufficient to wear away the working surface of the pump impeller.

Cavitation is typically caused by:

- A retreating water level, causing an unacceptable increase in the suction lift.

- A blockage in the suction line or in the footvalve.
- The footvalve resting on the dam floor or river bed.
- The footvalve being jammed shut.
- Too much friction in the suction line, caused by too long or too small diameter suction pipe.
- An air leak into the suction line.

Consequently, cavitation can be avoided by:

- Locating the pump site as close as possible to the water, mounting the pump on rails or similar to move it toward water as the water level retreats.
- Using a large diameter suction line (usually at least one size larger than the delivery pipe) and installing the suction line correctly to avoid leaks at joins.
- Locating the footvalve on a float to keep it off the bottom.
- Installing the suction line so that it rises continuously toward the pump and so the pump inlet flange is the highest point.
- Keeping the suction line short, with as few bends and fittings as possible.
- Installing a coarse screen on the suction line intake, to avoid large obstacles entering the pipe (a fine screen will get blocked too easily).

Follow the manufacturer's recommendations in regard to maximum suction lift for any particular pump.

Pump maintenance

Pumps do not normally require a high degree of maintenance. Follow the manufacturer's recommendations for bearing lubrication, if they are not a sealed type, using a water resistant grease at the required interval. Do not overgrease. Glands and seals may require occasional attention, particularly if the pump is inadvertently allowed to run dry. Pumps with an extended drive shaft will have bearings to support the shaft at fixed positions. These may be grease lubricated, sometimes from a common lubrication gallery, or may be a floating type that is water lubricated.

Most pump maintenance problems occur when pumping dirty water. The pump inlet should be screened to prevent entry of trash and larger debris. The screen should not be too fine or too small in size that it blocks too easily, since this could restrict water entry to the pump and cause cavitation. Self-cleaning inlet screens are available if the water supply is sufficiently dirty. Some pump types are better able to handle material carried with the water, but this material must be removed before delivery through irrigation pipes anyhow.

Abrasive particles may be present in the water supply, which may not be sufficiently large to cause a blockage, but can wear pump parts (and pipelines). If necessary, wear-resistant alloys can be specified, if the material cannot be removed. When pumping from dams, the pump inlet can be suspended from a float to access the cleanest water.

If the power unit is a diesel or petrol engine, normal maintenance procedures for these machines will be required. An hour-meter will assist maintenance scheduling, as well as irrigation record keeping.

The drive between pump and power unit (engine or electric motor) may also require maintenance. A shaft drive may have universal joints that require greasing, and lubrication maintenance will be necessary if the drive contains a gearbox. Belt drives will require periodic adjustment to belt tension, and occasional belt replacement. (Where multiple belts are used, replace all belts together.)

Ensure all safety guards are in place after maintenance has been conducted.

Pipelines

For irrigation applications, the design of pipelines is a specialised activity, requiring simultaneous consideration of a range of factors to provide an optimum design. This section summarises factors, where expert advice should be sought.

Pipelines can be permanently installed or be portable, buried or above ground. Pipes can be located to deviate around difficult terrain but the shortest distance is usually the least cost.

Pipe material

A range of pipe materials is available for irrigation use.

PVC (polyvinyl chloride) is in common use, available in a wide range of diameters and pressure ratings, and is supplied in 6 m lengths of pipe. Two methods are available for joining these pipes:

- Solvent weld, where a chemical is applied to the surface of each of the joints to partially soften the material. Contact between the two surfaces welds plastic to plastic.
- Rubber ring joint, where a special rubber ring provides a seal between the two pipe ends.

Both require correct joint preparation and assembly.

PE (polyethylene) is in widespread use for small diameter applications such as microirrigation laterals. It is also available in much larger diameters for high flow reticulation applications, in 12 m lengths, or coiled for diameters of up to

125 mm. In small diameters, it is joined by a bayonet (for low pressure) or screwed fitting.

Larger diameter pipes are joined by butt welding in the field using a special machine. This provides a basis for homogeneous (pipe to pipe) joints, butt flange joints and demountable joints using Victaulic couplings. Cutting threads in PVC pipes is not recommended.

Where PVC is installed in a trench, PE pipe may lie on the surface. A potential problem of above ground PE pipe, being black, is the amount of thermal expansion and contraction, when exposed to the sun, then subjected to cold water or night time temperatures. Its coefficient of expansion is typically 24 mm per 100 m of pipe for each 10°C change in temperature.

Various types of steel pipe are available, usually for special applications where plastic pipes lack sufficient strength, such as large diameter conduits, under roadways, suction

1. Cut pipe square, deburr, chamfer internal edge of pipe with knife.

2. Separate components of fitting and mount them on pipe, first the nut, followed by the split ring. Make sure the large end of the split ring faces towards the fitting.

3. Insert barbed end of tail into pipe so that flange is hard against pipe face. If necessary, use a rubber mallet or a piece of timber and hammer.

4. Insert spigot end of tail into body of fitting until it passes through rubber O-ring and flange butts against shoulder of fitting. Lubricate with water or Vaseline.

5. Push split ring hard against flange and firmly hand tighten nut onto fitting body.

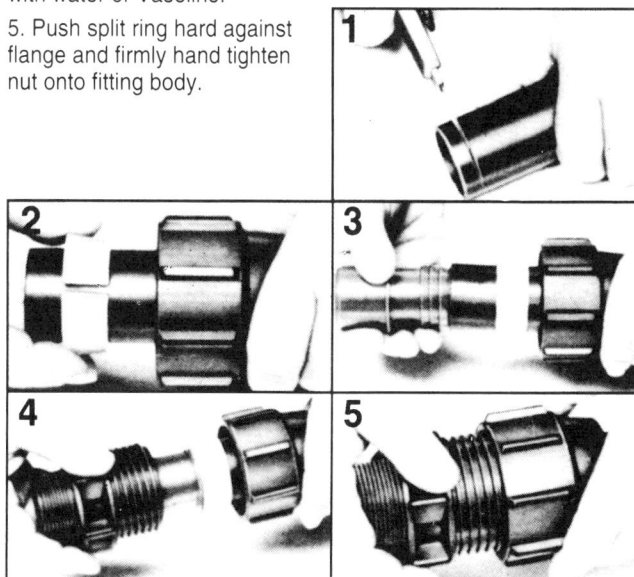

Figure 8.5 *Connection of small bore polythene pipes (Courtesy: Vinidex tubemakers P/L)*

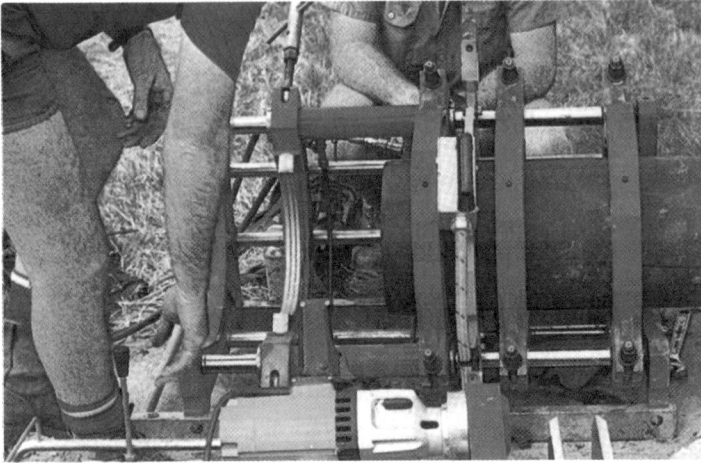

Figure 8.6 *Butt welding of large bore PE pipe on site. The finished pipe can be hauled into position, or the welding machine can be progressively moved from join to join (Courtesy: Vinidex Tubemakers P/L)*

applications. Cement or plastic coatings and linings are also available, and the pipe could be plain, galvanised or stainless steel, straight walled or corrugated.

Other pipe materials include fibre reinforced cement, aluminium (for hand shift spraylines and gated pipe), and reinforced synthetic lay-flat hose material.

Regardless of the type of pipe, ensure that it and its fittings are manufactured to the relevant Australian standards, preferably by a quality endorsed manufacturer.

Selection of pipe diameter

A number of factors influence this decision. Of prime importance is the amount of friction created between the water and the internal surface of the pipe, and that associated with water turbulence.

If the amount of friction is too high, the pump must operate at a higher pressure, wasting power, and there will be a larger difference in pressure between upstream and downstream sections of pipe.

The amount of friction is influenced partly by the roughness of the pipe material (plastic pipes are generally considered smooth walled), and its length, but more so by the water velocity. For a given flow rate, pre-determined by a previous estimation of peak irrigation requirements, the water velocity is determined by the pipe internal diameter. As a consequence, the pipe diameter is chosen to keep the water velocity to an acceptably low level.

Long lengths of pipe need careful consideration, since pressure losses from friction are proportional to pipe length. Losses caused by friction can be tolerated better if the system operates for fewer hours per season, since the higher operating costs are justified compared to the higher purchase price of a

Figure 8.7 *Flanges welded to pipes in the field ready for final positioning and bolting together (Courtesy: Vinidex Tubemakers P/L)*

Figure 8.9 *Demountable Victaulic coupling using shouldered ends welded onto the pipes (Courtesy: Vinidex Tubemakers P/L)*

larger diameter pipe. On the other hand, a pressure variation of only 10 per cent or so is normally specified for microirrigation and spray irrigation laterals, to ensure an acceptably uniform application rate.

The above discussion has not included the effect elevation has on water pressure. If water is running downhill, pressure is gained because of the effects of gravity, which offsets the pressure lost from friction. A pipe diameter can be selected where one closely balances the other, to give a uniform pressure distribution. However, when water runs uphill, pressure is lost from friction as well as elevation, making it impossible to achieve even pressure. As a general rule, it is often preferable to deliver water to the highest point of an irrigation block, then run the laterals or spraylines downhill.

There is an extra pressure loss associated with the turbulence caused by pipe fittings, and as water passes through valves, filters and other components. These pressure losses need to be considered in addition to pressure losses in straight lengths of pipe, but are also strongly dependant on water velocity; if water velocity is kept low, pressure losses from fittings will also be kept acceptably low. Head losses from all sources should be around 10 per cent of the total head.

A high water velocity will also result in high surge pressures associated with water hammer. Although the pipe might cope with normal operating pressures, sudden pressure surges (as a result of rapid valve closure, or sudden pump start-up or shut-down) can travel along the pipe, exceeding its pressure rating, or causing pipe movement. Repeated surge pressures can result in a fatigue failure of the pipe some time after installation. Ideally, water velocity should be below 2 m/sec.

In summary, optimum pipe diameter (for the selected pipe material) needs to consider a number of site specific factors: flow rates in each section of the pipe network and the variation in flow rates during operation of the system; overall running time of the system; pressure variations along each section of pipe, its fittings, and other irrigation components; low water velocity; differences in elevation throughout the pipe network; cost; availability of different pipe diameters. Balancing these factors requires careful analysis, not guesswork.

Pressure rating

For each type of pipe material, there is a maximum internal pressure that it is designed to withstand, as recommended by the manufacturer. Steel pipes can accommodate relatively high pressure, but plastic pipes need to be carefully selected for high pressure applications, particularly where surge pressures could be encountered. The pressure rating of standard class plastic pipe is 600 kPa (60 m head), but a wide range of alternative pressure ratings is available.

For plastic pipes, an increased pressure rating is provided by increasing the wall thickness of the pipe, by making the internal diameter smaller. This has two effects:

- increasing unit cost substantially,
- increasing water velocity at any particular flow rate.

Both of them are undesirable, so water pressure and pipe pressure rating need to be considered carefully. Pressure ratings are reduced at elevated temperature for plastic pipe materials.

Some pipes may be subjected to partial vacuum, such as pump suction lines, and occasionally other sections of the pipe network. Thin-walled pipes may have a high rating to internal water pressures, but may not be adequate for vacuum pressures which could cause the pipe wall to collapse inward. Where vacuum is unavoidable, the wall thickness should be selected accordingly, even if it is more than sufficient for internal water pressure. Other sections of pipe can be protected by vacuum relief devices, as required. These admit air into the pipeline, for example after pump shut-down, then close again when water pressure is provided.

External load may be applied to pipes under roadways, in deep trenches, or under high embankments. Pipe wall strength will need to consider this as well.

Pipe installation

Although spray and microirrigation laterals are above ground, most reticulation pipes will be buried below ground. Minimum standards apply to joining and installing pipes and associated components. Specific recommendations are available from the manufacturer, and in various Australian Standards, (for example AS 2032 , Installation of PVC Pipe Systems), but the following general comments are relevant.

Pipes can be joined by flanged or screwed couplings, special sealing rings or welding. Follow the manufacturer's recommendations for each type of pipe. For welded pipe, correct joint preparation is essential, whether it be steel or plastic pipe. For solvent weld joins in PVC, minimum standards apply to the cleaning of the two parts to be joined, the amount of solvent to be used (excess solvent around the edges of a completed join can continue to weaken the pipe adjacent to the join), and the amount of overlap of the two ends in the join.

Figure 8.9 *Installation of large bore PVC pipe. Trench width, bedding conditions, and backfilling technique should comply with manufacturers requirements. Rocky terrain such as this poses particular difficulties, as the pipe must be protected against rock contact*

Figure 8.10 *Concrete thrust blocking prior to completion of backfilling*

Rubber-ring joins use a specially shaped sealing ring that needs to fit one way only. Care should be taken when installing joined pipes in a trench, particularly for rubber-ring jointed pipe which can move out of alignment after initial fitting.

When installed in a trench, there should be sufficient clear space either side of the pipe, and a minimum depth of fill above the top of the pipe depending on the amount of traffic expected to pass over the pipe (450 mm is the normal minimum). Although normally permitted, check with the relevant authorities before using the same trench for other services such as electricity conductors. The floor of the trench should be smoothed, and the initial trench backfill should be of loose material free of stones. Care should be taken to tamp the fill beside the pipe, to ensure the pipe is fully supported in the trench.

Backfilling should be done in two stages, to minimise the amount of settlement in the trench after rainfall. Where the trench passes through stony ground, it should be partly backfilled with loose sand, imported if required, to protect the full perimeter of the pipe from damage from stones.

Pipe movement must be eliminated, particularly for rubber-ring jointed pipes, where the joints can come apart, but also for other pipes where excessive movement can crack joins. Concrete thrust blocks are essential at any change of direction for rubber ring jointed pipe, where the momentum of water hitting the fitting must be supported. Follow the manufacturer's recommendations for the size and arrangement of the thrust block, but note that it is not preferred to fully encase the pipeline in concrete, and the earth behind the concrete needs to be an undisturbed part of the excavation, not loose fill. All hydrant and stand pipes need to be supported to prevent accidental damage to the pipes they are connected to. Contraction of the pipe when first filled with cold water should be considered.

Following installation, with thrust blocks installed and cured, the pipe should be pressure tested. This is preferably done with the pipe secured but joins exposed, so leaks can be observed, but leaving trenches partially open for this is somewhat impractical.

Standard leak checking procedure involves pressurising the pipe to a certain percentage above its normal operating pressure, then holding that pressure for a period of time. If the pressure drops during the test period, it must be leaking somewhere (a very small pressure drop is permitted, to allow for pipe expansion and the effects of dissolved gases).

Jointing procedures for PVC pipes

The following pages, courtesy of Vinidex Tubemakers Pty Limited, describe the recommended procedures for joining PVC pipes.

INTRODUCTION

Two points critical to the solvent cement jointing procedure are:

- The solvent cement and priming fluid used should be produced in accordance with AS3879 (1991) — Solvent cements and priming fluid for use with uPVC pipes and fittings.

- Solvent cement jointing is a trade skill and should be executed only by appropriately qualified persons.

The PLASTEK *Manual Accreditation for Quality installation of uPVC Water and Sewer Pipe* states the following:

"Only the primer and solvent cemnent conforming to Interim AS3879 (1991) supplied by the pipe manufacturer should be used for jointing. Jointing primers and solvents available from other sources cannot be guaranteed to give the same high quality results" (PLASTEK Accreditation Program — Melbourne Water).

Vinidex premium solvent cements and priming fluid are produced in accordance with AS3879.

Vinidex recommends Vinidex solvent cements and priming fluid for use with Vinidex PVC pipes and fittings, thus ensuring a complete quality system. Vinidex premium solvent cements and priming fluid are specially formulated for PVC pipes and fittings and should **not** be used with other thermoplastic materials.

The following procedure should be strictly observed for best results. The steps and precautions will allow easy and efficient assembly of joints. Users may refer to Australian Standard 2032 'Installation of uPVC Pipe System' for further guidance.

Incorrect procedure and short cuts will lead to poor quality joints and possible system failure.

SOLVENT CEMENT JOINT PRINCIPLES

Priming Fluid, Type P & Type N Solvent Cement

There are two types of joint:

- Pressure — joint with an interference fit
- Non-pressure — joint that may have a small clearance

Sockets made to Australian Standards AS1477 for pressure pipes and fittings are tapered, ensuring the right level of interference. This may not apply to all pipes and fittings, particularly from other countries.

Vinidex offers two types of solvent cements formulated specifically for pressure and non-pressure applications. They are colour coded, along with the primer in accordance with AS3879:

- Type P for pressure, including potable water installations, designed to develop high shear strengths with an interference fit, (green solvent, green print and lid).

- Type N for non-pressure applications, designed for the higher gap filling properties needed with clearance fits, (blue solvent, blue label and lid).

- Priming fluid for use with both solvent cements, (red priming fluid, red label and lid).

Always use the correct solvent cement for the application.

Source: Vinidex Tubemakers Pty Limited, *1989 The Water Supply Manual for PVC Pipe Systems*

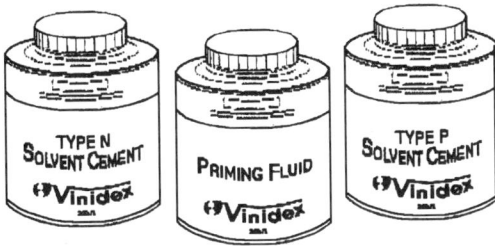

Solvent cement jointing is a "chemical welding", not a gluing process. The priming fluid cleans, degreases and removes the glazed surface thus preparing and softening the surface of the pipe so that the solvent cement bonds the PVC.

The solvent cement softens, swells and dissolves the spigot and socket surfaces, that bond into one solid material as they cure.

PROCEDURE

1. Prepare the Pipe

Before jointing, check that the pipe has been cut square and all the burrs are removed from the inside and outside edge. Remove the sharp edge from the outside and inside of the pipe with a deburring tool. Do not create a large chamfer that will trap a pool of solvent cement. Remove all dirt, swarf, and moisture from spigot and socket.

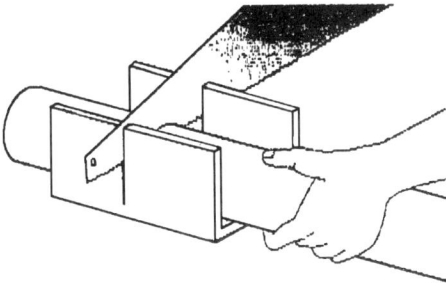

2. Witness Mark the Pipe

It is essential to be able to determine when the spigot is fully home in the socket. Mark the spigot with a pencil line ('witness mark') at a distance equal to the internal depth of the socket. Other marking methods may be used provided that they do not damage or score the pipe.

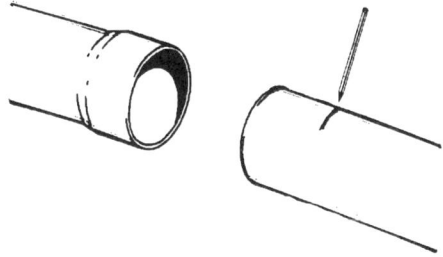

3.'Dry Fit' the Joint

'Dry fit' the spigot into the socket, check the pipe for proper alignment. Any adjustments for the correct fit can be made now, not later. For pressure pipes, the spigot should interfere in the socket before it is fully inserted to the pencil line. Ovality in the pipe and socket will automatically be re-rounded in the final solvent cementing process, but heavy walled pipe may give a false indication of the point of interference. Do not attempt to make a . pressure pipe joint that does not have an interference fit. Contact Vinidex if this occurs.

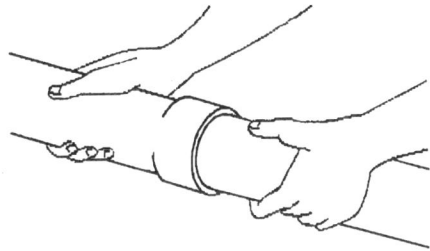

4. Prepare with Priming Fluid

Dre, degrease and prime the spigot and socket with a lint free cloth (natural fibres) dampened with Vinidex priming fluid. Priming is vitally important, as it etches off the gloss from the PVC, it softens the PVC surface for the solvent cement's effective bond. Use

---

Source: Vinidex Tubemakers Pty Limited, *1989 The Water Supply Manual for PVC Pipe Systems*

protective polyethylene gloves. Vinidex priming fluids are to be used before solvent cementing, prime and solvent cement one joint at a time.

5. Brush Selection

The brush should be large enough to apply the solvent cement to the joint in a maximum of 30 seconds. Approximately one third the pipe diameter is a good guide. Do not use the brush attached to the lid for pipes over 100mm in diameter. Decanting is not advisable, and excess should never be returned to the can. For large diameter pipes, it may be necessary to decant to an open larger vessel for a large brush to be used, in this case decant for one joint at a time.

Table of Recommended Brush Selection

Diameter size of pipe (mm)	Recommended size of brush (mm)
15,20,25,32,40,50	use brush supplied
65,80	25
100,125	38
150	50
200	83
225,250	75
300,375	100

6. Apply Solvent Cement

Using a suitably sized brush, apply a thin even coat of solvent cement to the internal surface of the socket first. Solvents will

evaporte faster from the exposed spigot than from the socket. Special care should be taken to ensure that excess solvent cement isn't built up at the back of the socket (pools of solvent will continue to attack the PVC and weaken the pipe). Then apply a heavier, even coat of solvent cement up to the witness mark on the spigot. Ensure the entire surface is covered. A 'dry' patch will not develop a proper bond, even if the mating surface is covered. An unlubricated patch may also make it difficult to obtain full insertion.

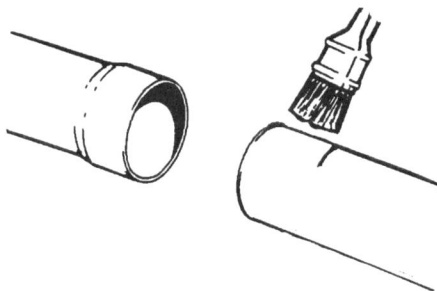

7. Inserting the Spigot

Make the joint immediately, in a single movement. Do not stop halfway, since the bond will start to set immediately and it will be almost impossible to insert further. It will aid distribution of the solvent cement to twist the spigot into the socket so that it rotates about a 1/4 turn whilst (not after) inserting, but where this cannot be done, particular attention should be paid to uniform solvent application.

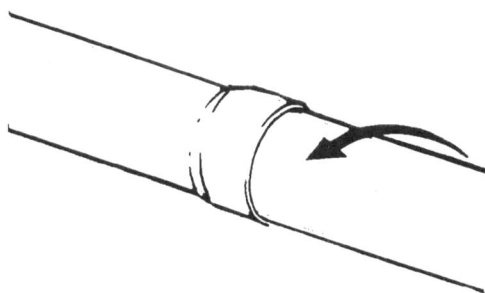

8. Push the Spigot Home

The spigot must be fully homed the full depth of the socket. The final 10 per cent of spigot

---

Source: Vinidex Tubemakers Pty Limited, *1989 The Water Supply Manual for PVC Pipe Systems*

penetration is vital to the interference fit. Mechanical force will be required for larger joints. Be ready in advance. Pipe pullars are commercially available for this purpose. Polyester pipe slings are very useful for gripping a pipe, in order to apply a winch or lever.

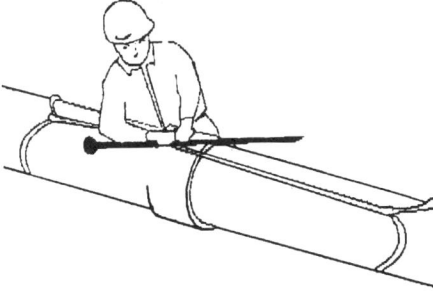

9. Hold the Joint

Hold the joint against movement and rejection of the spigot for a minimum of 30 seconds. Disturbing the joint during this phase will seriously impair the strength of the joint.

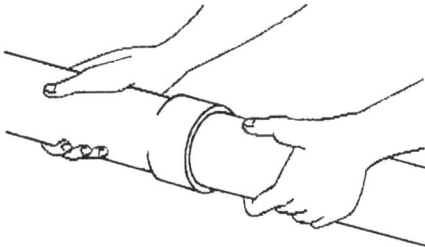

10. Wipe off Excess Solvent Cement

For a neat professional joint, with a clean rag wipe off excess solvent cement immediately from the outside of the joint.

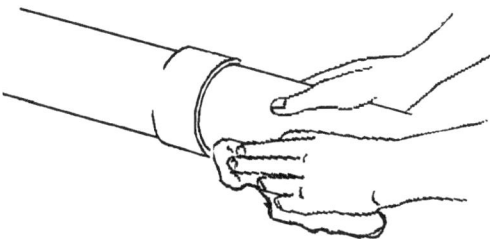

11.Do Not Disturb the Joint

Once the joint is made, do not disturb it for five minutes or rough handle it for at least one hour. Do not fill the pipe with water for at least one hour after making the last joint. Do not pressurise the line until fully cured.

12.Cure the Joint

The process of curing, is a function of temperature, humidity and time. Joints cure faster when the humidity is low and the temperature is high. The higher the temperature the faster joints will cure. As a guide, at a temperature of 16°C and above, 24 hours should be allowed, at 0°C, 48 hours is necessary.

PRECAUTIONS TO ACHIEVE AN EFFECTIVE JOINT

To achieve strong sealed, leak free and safe jointing, and significant long and short term installation savings, these additional precautions should be followed:

Cutting and Joining

a) Make sure that the end of each pipe is square in its socket and in the same alignment and grade as the preceding pipes or fittings. Cut the pipe using a fine toothed saw and mitre box or circular saw with an abrasive disc. To ensure full interference fit, the last few millimetres of spigot count.

b) Create a 0.5mm chamfer, as a sharp edge on the spigot will wipe off the solvent and reduce the interface area. Remove all swarf and burrs so that later filings cannot become dislodged and jam taps and valves.

c) Do not attempt to joint pipes at an angle. Curved lines should be jointed without stress, then curved after the joint is cured. Support the spigot and socket clear of the ground when jointing, this will avoid contamination with sand or soil.

d) An unsatisfactory solvent cement joint cannot be re-executed, nor can previously cemented spigots and sockets be re-used. To effect repairs, cut out the joint and remake or use machanical repair fittings.

Apply Correct Solvent Quantity

The correct amount of solvent is a uniform self levelling layer without runs, achieved by experience and judgement.

Too much solvent will form pools and continue to attack and weaken the pipe. Too little solvent will require you to brush out excessively, the solvent will quickly evaporate with vigorous brushing.

Take care not to spill solvent cement onto pipes or fittings. Accidental spillage should be wiped off immediately.

Open Time

Vinidex Type P and N Solvent cements satisfy the long term pressure test procedure of AS3879 requiring an open time of 3 minutes. Open time is the time from the beginning of solvent application until the jointing of the parts.

Important: In the field, allowable open time can vary considerably because weather conditions can influence the drying time of solvent cements. Each joint should be completed immediately.

Adverse Weather

High temperatures and air movement will radically increase the loss of solvents, and solvent cement jointing should not be performed when the temperature is more than 35°C. Some form of protection should be provided when jointing in windy and dusty conditions.

When jointing under wet and very cold conditions, make sure that the mating surfaces are dry and free from ice, as moisture may prevent the solvent cement from obtaining its maximum strength.

Storage

Keep the containers stored below 30°C. The solvent cement lids should be tightly sealed when not in use to prevent evaporation of the solvent. Do not use solvent cement that has gone cloudy or has started to gel in the can. Do not use solvent cement after the "use by" date shown on the can, the chemical constituents can change over a long period of time, even in a sealed can.

Health

Vinidex solvent cement and priming fluid have been specially formulated for jointing Vinidex PVC pipe. Forced ventilation should be used in confined spaces. Do not bring a naked flame within the vicinity of solvent cement operations.

Spillage onto the skin should be washed off immediately with soap and water. Should the solvent cement get in your eyes, wash them with clean water for at least 15 minutes and seek medical advice.

If solvent cement or priming fluid is swallowed, do not induce vomiting, dilute by giving two glasses of water, and seek medical advice immediately.

Average Number of Joints per 500 ml

The following table provides and indication as to the number of joiints that are made per 500ml container of Priming Fluid and Solvent Cement. (*Table on next page*)

Source: Vinidex Tubemakers Pty Limited, *1989 The Water Supply Manual for PVC Pipe Systems*

Size of Pipe DN (mm)	Priming Fluid	Solvent Cement
15	1050	300
20	625	175
25	450	130
32	325	95
40	250	70
50	150	42
65	125	35
80	100	30
100	70	25
125	60	20
150	45	15
200	25	10
225	15	7
250	12	6
300	12	5
375	12	5

PRODUCT DATA

Nominal* size (mm)	Mean O.D. (mm)	PN	Mean Bore (mm)	Mean Wall (mm)	Approx Mass kg/6m
100	121.9	12	108.5	6.7	23
		16	104.3	8.8	29
		18	100.3	9.8	32
		20	100.3	10.8	35
150	177.4	12	157.8	9.8	48
		16	152.0	12.7	61
		18	149.0	14.2	67
		20	146.0	15.7	73
200	232.3	12	209.2	11.5	74
		16	202.2	15.0	95
		18	198.6	16.8	105
		20	195.2	18.5	115
225	257.3	12	233.7	12.8	93
		16	225.7	16.8	119
		18	221.9	18.7	132
		20	218.1	20.6	144
250	286.2	12	258.0	14.1	112
		16	249.2	18.5	145
		18	245.0	20.6	160
		20	240.8	22.7	174
300	345.4	12	311.4	17.0	164
		16	300.8	22.3	213
375	426.2	6	404.6	10.8	130
		9	394.2	16.0	189
		12	384.4	20.9	246
		16	371.2	27.5	319
450	507.0	6	481.4	12.8	#
		9	469.2	18.9	#
		12	457.2	24.9	#
		16	441.6	32.7	#

Not all classes and sizes are available from Vinidex locations. Contact your nearest Vinidex office for latest information of the Vinyl iron range.

 # Contact your nearest Vinidex office

Source: Vinidex Tubemakers Pty Limited, *1989 The Water Supply Manual for PVC Pipe Systems*

PIPE CLASSIFICATION _ PRESSURE RATINGS

AS1477 classifies Vinyl iron pipes into the pressure classes shown in the table below. The pipe classification is for a maximum static working pressure at 20°C pipe material temperature.

PN		6	9	12	16	18	20
Metres Head		61	91	122	163	184	204
Mpa		0.6	0.9	1.2	1.6	1.8	2.0

Maximum working pressure at 20°C pipe material pressure

SELECTION OF CLASS

The above table is intended to provide a first order guide to the duty for which the pipes are intended. These working pressures incorporate a suitable factor of safety to ensure trouble free operation under average service conditions.

For specification purposes, the design point adopted for PVC pipes is the 50 year regression line with a factor of safety of 2.14. At 100 years the factor of safety is 2.08.

There are many factors which must be considered when determining the severity of service and the appropriate class of pipe.

The final choice is up to the designer in the light of knowledge of the particular situation.

Amongst the factors to be considered are:

operating pressure characteristics, temperature, external loading, service life required and factor of safety.

For more detailed information refer to the Vinidex publications: *The Water Supply Manual for PVC Pipe Systems* and *Flow Charts for PVC Pressure Pipe.*

HANDLING AND STORAGE

Genersl rules for handling, storage, and installation of PVC pipes given in Australian Standard 2032 should be observed.

Local authority regulations and specifications take precedence over all other instructions.

PVC pipe is very robust, but still can be damaged by rough handling. Pipes should not be thrown from truck or dragged over rough surfaces. Care should be taken when handling PVC pipes in low temperatures.

Since the soundness of any pipe joint depends on the condition of the spigot and socket, special care should be taken not to allow them to come into contact with sharp edges or protruding nails.

Transportation of PVC Pipes

While in transit, pipes should be well secured and supported. Chains or wire ropes may be used only if suitably padded to protect the pipe from damage. Care should be taken that the pipes are firmly tied so that the sockets cannot rub together.

Pipes may be unloaded from vehicle by rolling them gently down timbers, care being taken to ensure that the pipes do not fall onto one another or onto any hard or eneven surface.

Storage of PVC Pipes

Pipes should be given adequate support at all times. Pipes should be stacked in layers with sockets placed at alternate ends of the stack and with the sockets protruding.

Horizontal supports of about 75mm wide should be spaced not more than 1.5m centre-to-centre beneath the pipes to provide even support.

Vertical side supports should also be provided at intervals of 3m along rectangular pipe stacks.

For long term storage (longer than 3 months) the maximum free height should not exceed 1.5m The heaviest pipes should be on the bottom.

Crated pipes, however, may be stacked higher provided that the load bearing is not

---

Source: Vinidex Tubemakers Pty Limited, *1989 The Water Supply Manual for PVC Pipe Systems*

taken directly by the lower pipes. In all cases, stacking should be such that the pipes will not become distorted.

If it is planned to store pipes in direct sunlight for a period in excess of one year, the pipes should be covered with material such as hessian. Coverings such as black plastic must not be used as these greatly increase the temperatures within the stack.

JOINTING SEQUENCE

- Clean the socket, especially the ring groove. Do not use a rag with lubricant on it.

- Check that the spigot end, if cut in the field, has a chamfer of approximately 12° to 15°. When a pipe is cut, a witness mark should be pencilled in and care should be taken to mark the correct position in accordance with the table on **page 5**.

- Insert the rubber ring into the groove with the colour marking on the ring facing outwards. The rubber ring is correctly fitted when the thickest cross section of the ring is positioned towards the outside of the socket and the groove in the rubber ring is positioned inside the socket.

  Run your finger around the lead-in angle of the rubber ring to check that it is correctly seated, not twisted, and that it is evenly distributed around the ring groove.

- Clean the spigot end of the pipe as far back as the witness mark. Apply Vinidex jointing lubricant to the spigot end as far back as the witness mark and especially to the chamfered section.

  Keep the rubber ring and ring groove free of jointing lubricant until the joint is actually being made.

- Align the spigot with the socket and apply a firm, even thrust to push the spigot into the socket. It is possible to joint 100mm and 150mm diameter joints by hand. Larger diameter pipes may require the use of a

Source: Vinidex Tubemakers Pty Limited, *1989 The Water Supply Manual for PVC Pipe Systems*

bar and timber block. Alternatively, a commercially available pipe puller may be used to joint the pipes. Brace the socker end of the line so that proviously jointed pipes are prevented from closing up.

- Inspect each joint to ensure that the witness mark is just visible at the face of each socket.

  Important: Pipe joints must not be pushed home to the bottom of the socket. They must go no further than the witness mark. This is to allow for possible sxpansion of the pipe.

Nominal Size (mm)	Approx. Length of Chamfer Lc (mm)	Witness Mark Lw (mm)
100	12	103
150	13	122
200	21	193
225	28	203
250	23	204
300	32	228
375	40	238
450	*	*

* Contact your Vinidex Office

Vinyl Iron Spigot Dimensions

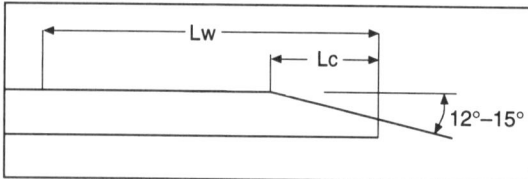

Pipe Chamfer

NOTES ON JOINTING

Pipes may be jointed out of the trench but it is preferable that connections be made in the trench to prevent possible "pulling" of the joint.

Always joint the pipe in a straight line and then lay to a curve if required.

If a pipe joint is homed too far, it may be withdrawn immediately, but once the lubricant is dry (which takes only a few minutes in hot weather) machanical aids are required to pull the joint apart.

If excessive force is required to make a joint, this may mean that the rubber ring has been displaced. To check placement of the ring without having to dismantle the joint, a feeler gauge can be inserted between the socket and pipe to check even placement of the ring.

Vinyl Iron PVC pipe can be inserted into the standard socket of a Ductile Iron pipe. However, caution should be taken if inserting a Ductile Iron pipe spigot into a Vinyl Iron PVC pipe socket. Ductile Iron & Vinyl Iron pipes are manufactured to different tolerances. The wider tolerance of Ductile Iron pipes may adversely affect joint integrity if a Vinyl Iron socket is damaged by a dimensionally incompatible Ductile Iron pipe spigot.

CUTTING

During manufacture, pipes are cut to standard length by cutoff saws. These saws have carbide tipped circular blades which produce a neat cut without burrs.

However, pipes may be cut on sit4 with a varity of cutting tools. These are:

1. Proprietary cutting tools.

   These tools can cut, deburr and chamfer the pipe in one operation. They are the best tools for cutting pipe.

2. A portable electric saw with cutoff wheel.

   This is quick and easy to use and produces a neat clean cut requiring little deburring. It does, however, require a power supply and the operator has to be skilled in using it to produce a square cut.

3. A hand saw and mitre box.

Source: Vinidex Tubemakers Pty Limited, *1989 The Water Supply Manual for PVC Pipe Systems*

This saw produces a square cut but requires more deburring. It takes comparatively more time and effort and requires a stand.

The use of roller cutters is not recommended.

Note: If the cut pipe end is not square, uneven pressure on the ring may cause difficulty in jointing.

CHAMFERING

If the pipe is to be used for making a rubber ring joint, a chamfer is required. Special chamfering tools are available for this purpose, but in absence of this equipment, a body file can be used provided it does not leave any sharp edges which may cut the rubber ring. Do not make an excessively sharp edge at the rim of the bore and do not chip or break this edge. The diagram on page 140 gives the length of chamfer required at 12° to 15° angle.

Make a witness mark at the appropriate distance from the end of the spigot, also as shown in the table.

JOINTING FITTINGS

Vinyl Iron pipes may be jointed to cast iron fittings to Australian Standard 2544. In this event the appropriate sealing rings and jointing prodedure recommended by the manufacturer of the fitting should be used.

Pipes should be jointed to the full depth of engagement available on the cast iron fitting socket. In general it will not be possible to align the witness mark to the end of the socket.

To ensure full engagement, check the depth of the fitting socket and mark this length on the spigot of the pipe with an indelible marking pen.

RUBBER RINGS

Jointing rings are supplied with the pipe and are embossed with 'Vinidex Vinyl Iron'. Each ring has a painted mark on its front edge. This mark must face out of the socket when the ring is inserted.

Rubber rings are manufactured and tested in accordance with AS1646 "Elastomeric Seals for Waterworks Purposes". Unless otherwise specified, natural rubber will be supplied for pressure pipe. Depending on the particular specification, the rubber used is either natural rubber (white dot), Styrene Butadiene rubber (SBR- blue dot) or Polychloroprene (Neoprene-red dot).

Warning: Rings marked with two coloured dots are sewer rings containing a chemical root inhibitor. They may also vary dimensionally from pressure rings and should not be interchanged.

JOINTING LUBRICANT

Vinidex jointing lubricant is supplied with the pipe and is a specially formulated organic preparation for easy jointing of Vinyl Iron pipe. The use of petroleum based greases or other substitutes may affect the ring or potability of the water supply and cannot be recommended. The approximate number of joints that may be jointed with one litre is as follows

Nominal Size (mm)	Number of Joints per litre
100	100
150	60
200	50
225	45
250	40
300	30
375	25
450	20

Joints per litre of lubricant

Source: Vinidex Tubemakers Pty Limited, *1989 The Water Supply Manual for PVC Pipe Systems*

INSTALLING PIPES ON A CURVE

Vinyl Iron pipes are sufficiently flexible to allow bending around a curved alignment or around obstacles. The same property enables the pipes to absorb subsidence and other earth movements.

Significant bending moments should not be imposed on joints, since this introduces undesirable stresses in the spigot and socket which may be detrimental to long term performance. To avoid this, the joints in curved lines must be thoroughly supported by compacted soil, with the bending occurring primarily at the centre of each pipe.

When installing pipes on a curve, the pipe should be jointed straight and then laid to the curve. The minimum bending radius of Vinyl Iron pipe is 300 times the diameter.

Nominal Pipe Size (mm)	Minimum Radius (mm)
100	30
150	45
200	60
225	67
250	75
300	90
375	112
450	135

Minimum radius of curvature

TAPPING

Only tapping saddles designed for use with PVC pipe should be used. These saddles should:

- Be contoured to fit around the pipe and not have lugs or sharp edges that dig in.
- Have a positive stop to avoid over tightening of the saddle around the pipe.

The maximum hole size that should be drilled in a PVC pipe for tapping purposes is 50mm or 1/3 the pipe diameter, which ever is smaller.

This does not prevent the connection of larger branch lines via tapping saddles, provided the hydraulic loss through the restricted hole size is acceptable.

For larger branches generally, a tee is preferred.

Holes should not be drilled into PVC pipe:

- Less than 300mm from a spigot end.
- Closer than 450mm to another hole on a common parallel line.
- Where significant bending stress is applied to the pipe.

Source: Vinidex Tubemakers Pty Limited, *1989 The Water Supply Manual for PVC Pipe Systems*

Channels

Channels are usually formed with banks made of earth, with a trapezoidal cross-section matched to the bed slope and the required flow rate, allowing some freeboard between high water level and the top of the bank. They are in common use for surface irrigation, using a variety of methods to get water from the channel onto the field, and some systems pump water from the channel into reticulation pipes or spraylines.

Channels take up more space than pipelines, which will be a disadvantage if the planting area is limited, or land area highly valuable. They are also difficult to cross, possibly limiting access between blocks and around the farm, and need higher maintenance compared to properly installed pipelines. Seepage can occur through the wetted perimeter of the bank resulting in inefficient water distribution and contributing to recharge to the water table. Where excessive and/or localised, the cost of lining the channel could be justified.

Channel capacity

The flow rate along an open channel can be estimated mathematically, and depends on the following factors:

- The cross-section and wetted perimeter of the channel.
- The bed slope along the channel.
- A "roughness coefficient", which is a measure of the resistance to flow at the surface of the channel.
- The restriction and subsequent change in water levels created at structures within the channel (such as gates and culverts).

The first two of these can be measured quite accurately, and although the roughness coefficient appears a bit vague, it can be reasonably estimated depending on whether the channel surface is bare earth or short maintained vegetation. Weed growth (and other debris) increase resistance to flow, and therefore reduce channel capacity. The location of the channel influences the average bed slope, although the bed elevation can be adjusted up (by building a pad) and down (by excavation below natural surface) to suit the local terrain. The capacity is largely determined by the channel cross section, principally its width.

Channel structures

Several types of structures may be required in a channel system:

- Gates, to divert water to channel branches, or to control channel water level or flow rate.

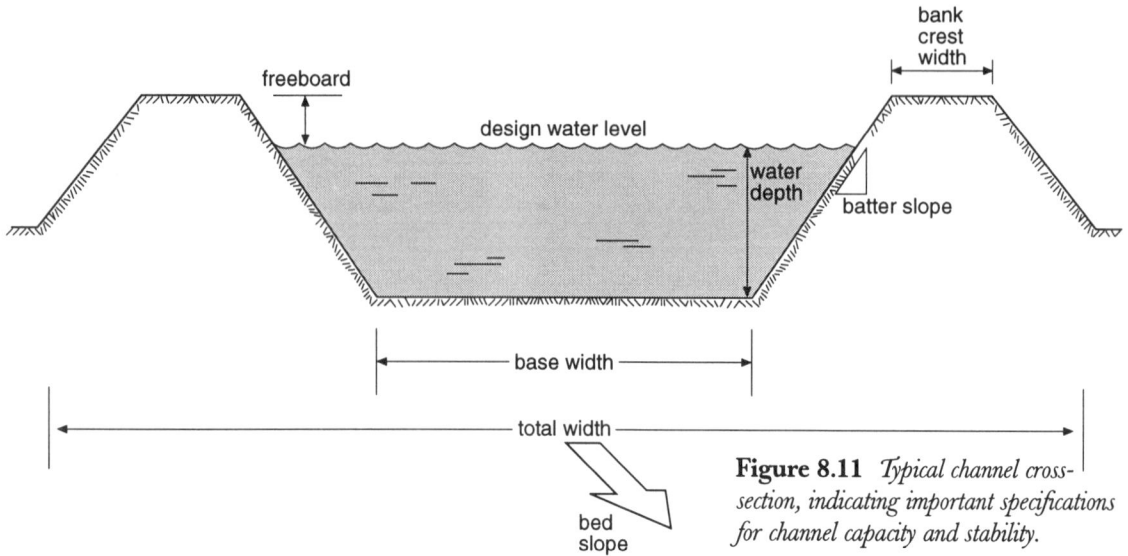

Figure 8.11 *Typical channel cross-section, indicating important specifications for channel capacity and stability.*

- Checks, to control upstream water levels.
- Drop structures, to lower the bed of the channel.
- Culverts, to provide a crossing point for roadways.
- Inverted siphons, to take water under a feature such as a road or floodway.
- Flowmeter, such as a Dethridge wheel (Figure 8.14).

Correct design is required for each, to ensure correct water levels and flow rates. A head loss occurs at each structure, which is reflected in a drop in water level, and this must be managed to ensure an adequate flow rate throughout the channel system.

Correct installation is necessary to avoid problems associated with localised scouring of the channel.

Figure 8.12 *Channel gates and culverts to direct flow and provide vehicle crossings. Gate size is selected to permit adequate flow rate and minimum head loss. Opening mechanisms can be mechanically assisted for large gates.*

Figure 8.13 *This structure takes water under a natural floodway*

Figure 8.14 *The Dethridge wheel meters the volume of water delivered to the farm*

Channel outlets

Alternative methods to get the water from the channel to the field include:

- Syphons: Small diameter, hand filled syphons are in common use, but because of the high labour requirement, large diameter, pump assisted units are increasingly popular.
- Gates located in the channel bank.
- Piped outlets.

Flow rate through the outlet will depend on the cross-section of the outlet, and the difference between water level in the channel and water level in the bay or furrow.

The flow velocity leaving the outlet can often be excessive, causing scouring of the earth in the vicinity of the outlet. If severe, silt deposited down the slope can interfere with correct water movement, and the scoured area will require repair.

Figure 8.15 *Some minor scouring of the earth is possible where high velocity flow is involved*

Channel maintenance

Channel maintenance may involve occasional re-shaping of the channel cross-section as a result of the gradual erosion of the earthen bed or walls, or from a build up of silt. Scouring damage at the inlet or outlet of channel structures should be monitored, and is best repaired or maintained using hard materials such as rock lining.

Vegetation growth in the channel can by managed by chemicals or by grazing, provided stock traffic does not damage the channel banks or cause low spots in the channel bed.

CHAPTER 9

Drainage

As a general rule, the optimum soil moisture content for plant growth is field capacity. This is the moisture content which occurs after a short period of drainage after thorough wetting, and whilst ensuring plant roots have ample access to soil moisture, there is also sufficient pore space for oxygen. (For a more detailed discussion of soil moisture, refer to Chapter 2).

Moisture in excess of field capacity can limit plant growth (and therefore potential yield) by reducing the availability of oxygen in the root zone. For some plants, this can be experienced after only a short duration of excess moisture, although some plants are more tolerant. A decision may need to be made whether to protect the crop or pasture by the construction of a drainage system. Relative tolerance to excess moisture (as measured by the yield reduction it causes) is only one of the factors which will influence whether to install a drainage system:

- The frequency of occurrence of excess moisture, and whether the crop is annual or perennial.
- The source of excessive moisture (rainfall on the site, runoff from higher ground, flooding from major watercourses, runoff from neighbours' properties or from culverts under roads, or a rising watertable). This will partly decide the best method of treatment.
- The suitability of the soil to accommodate drainage (in particular, the ability of moisture to move through and out of the soil profile into the drainage system).
- The location and suitability of a suitable site to dispose of drainage water once it has been collected.
- The quality of the drainage water (in particular the level of dissolved salt).

- The value of the extra production achieved after installing drainage, compared to the cost of installing and operating it.

A chapter on drainage is quite relevant to a book on irrigation, because both are concerned with the management of soil moisture content. Irrigation is a high cost activity often applied to high value crops, and the full benefit will not be achieved if there are occasional periods of excess moisture. Climatic patterns may be such that the site experiences annual wet and dry cycles. Under these circumstances, irrigation is necessary to ensure production during the dry, and drainage is necessary for the same reason in the wet. The processes that govern water movement through soil are applicable to both irrigation and drainage.

Sometimes, a drainage system can be designed so that surface runoff can be collected after rainfall on an irrigation area, and thereby contribute to the water supply for the farm. Occasionally, and contrary to best practice, the irrigation system applies too much water, or applies it inefficiently. The excess water can be collected by drainage.

Benefits

The primary purpose of drainage is to remove excess moisture from the root zone of plants, or to stop it reaching there.

A lack of oxygen in the root zone will inhibit root development, and therefore plant development, and possibly yield. This can be severe, even if the waterlogging is temporary, and can result in death of the plant if severe and/or prolonged.

The root system will not develop below the watertable, so if the watertable is higher than the maximum depth plant roots can penetrate,root growth will be reduced. Even if the watertable falls later in the season, development of the plant may be permanently affected. Good root development during periods of adequate moisture, as a result of drainage, can improve the plants' access to moisture during dry periods.

In extreme cases, the watertable is close to, or even above, the ground surface, so effective plant growth is not possible without drainage.

Excess moisture and a lack of oxygen will also affect the presence and activity of micro- and macro-organisms in the root zone, many of which are beneficial to soil quality and plants. A lack of which may be one of the reasons why a soil affected by waterlogging is sometimes described as "sour".

Wet conditions can greatly interfere with management practices on the farm, including machine operations such as

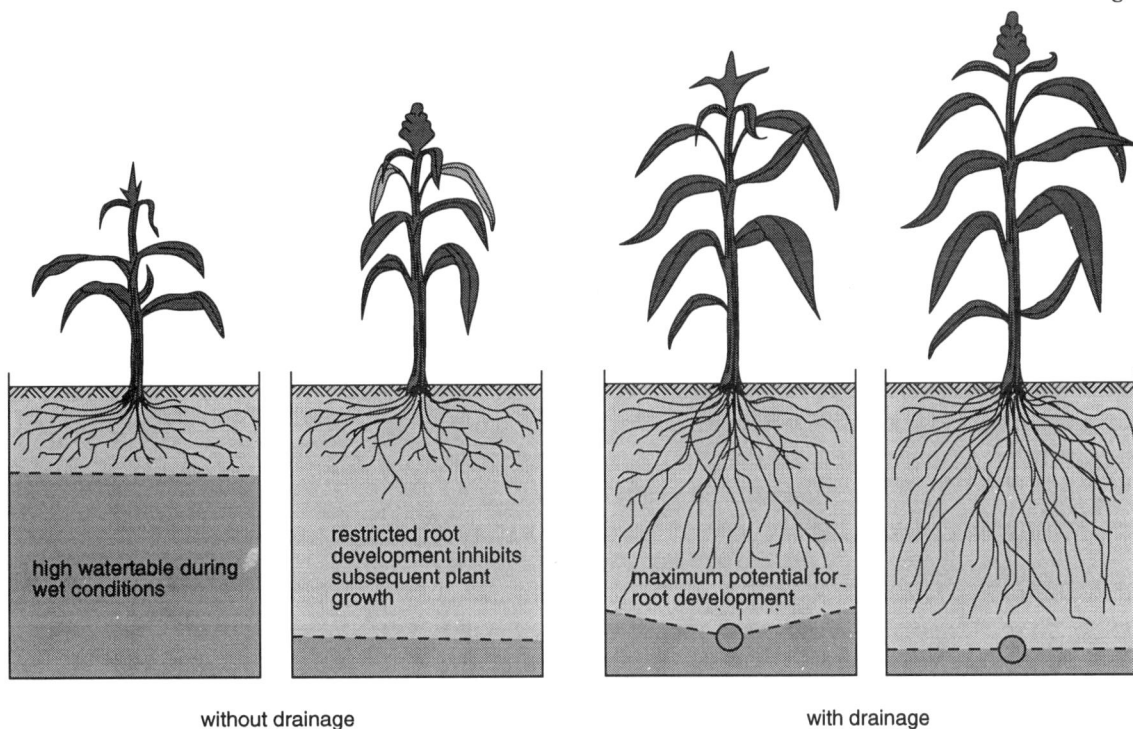

Figure 9.1 *The presence of a high watertable can seriously affect plant growth.*

Labels within figure:
high watertable during wet conditions

restricted root development inhibits subsequent plant growth

maximum potential for root development

without drainage

with drainage

cultivation, spraying and harvesting. Good drainage can contribute significantly to the timeliness of these operations, and hence to the overall profitability of the enterprise. Good trafficability will enhance general movement of stock, fodder and personnel around the farm, helping to keep routines on schedule. Wet conditions can contribute to stock ailments (particularly foot problems), pugging of soil under stock traffic, and some plant diseases. Weed and insect populations can become dominant, or at least, more difficult to manage. Cultivation when too wet (or too dry) can accelerate soil structural problems, and cause inferior results as well as excessive soil smearing and compaction.

Drainage can also play an important role in salinity management, by controlling the rise of saline groundwater into the root zone. Subsurface drains can be installed to intercept the water, maintaining watertable depth at an acceptable level. This water still needs to be collected at a central collection pond, ditch or sump, and disposed of safely (for example, to an evaporation area). Sometimes, salts accumulated in the root zone, or present in the irrigation water, can be leached into the drainage system. This can keep salt concentration to an acceptable low level in the root zone, provided the drainage water can be disposed of in an acceptable manner. Because water can move upwards from a stationary watertable, due to capillary action depending on soil texture, the safe watertable depth under saline conditions needs to be decided carefully.

Types of drains

Surface drainage refers to the use of open ditches, whereas subsurface drainage refers (usually) to the use of buried perforated pipelines (a variation on this is a mole drain, where an unlined drainage tube is constructed under the ground surface with a purpose built implement). Both types can be used to intercept water from higher ground, and in the control of watertable height, although not always equally effectively. Surface drainage is better able to control surface runoff.

The open ditches constructed for surface drainage can have some disadvantages. They can be expensive to install (although subsurface drainage is also expensive), require maintenance for controlling weed growth and siltation, they use ground which might otherwise be put under crop, and can create access difficulties for machinery and movement of stock.

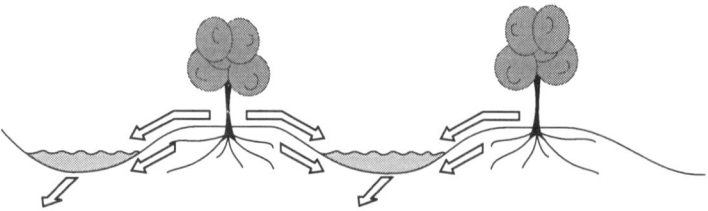

Figure 9.2 *Planting on raised beds can assist in removing surplus water, provided there is a slight gradient along the furrow or wheel track to encourage water to drain from the site. Water will more readily run off the surface of the bed, and seepage from the bed can also occur. Bed width (and therefore drain spacing) will depend on soil hydraulic characteristics, as well as machine width.*

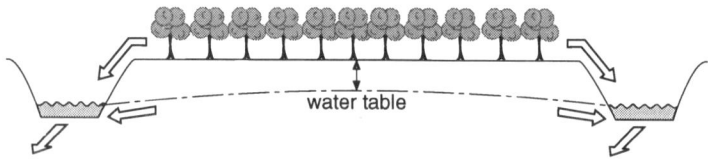

water table

Figure 9.3 *Ditches can be installed at wider spacings for some situations. Some control over watertable depth is achieved between the ditches.*

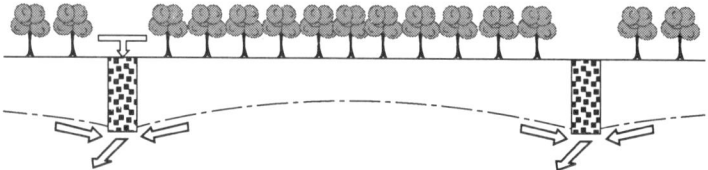

Figure 9.4 *Where ditches are inconvenient, and the affected area relatively small, gravel drains could be considered.*

Figure 9.5 *Perforated drainage pipe is buried in trenches, depth and spacing between them governed by site specific factors. Gravel packing may be necessary.*

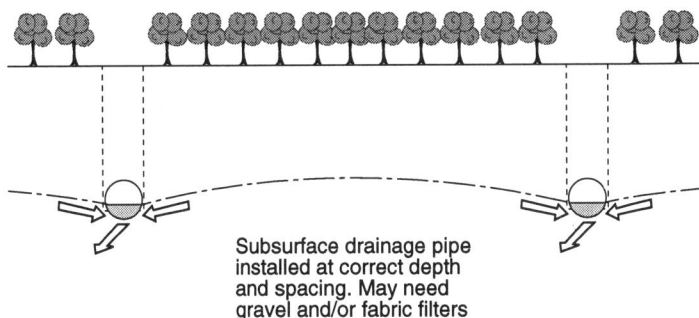

Subsurface drainage pipe installed at correct depth and spacing. May need gravel and/or fabric filters

Surface drainage can be particularly effective for localised surface ponding, or the control of runoff, particularly when integrated with a whole property plan. They are not always useful for watertable control, particularly in heavier soils, because the slow rate of water movement through the soil requires them to be too close together to be practicable. Surface drains are often used to collect tailwater and rainfall runoff from irrigated fields, and as the collection point for the discharge of subsurface drains.

The flow of water along a surface drain depends on its cross-sectional area, and bed slope. Their correct design therefore requires detailed analysis of the site topography, the drainage requirement, and the discharge conditions.

Subsurface drains have the potential to provide greater control over watertable depth, with minimum disturbance to surface conditions (following installation). Surface ponding can also be partly controlled by providing for the entry of surface water, as shown in Figure 9.7.

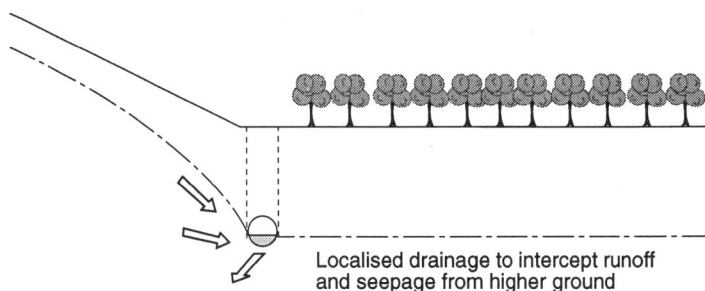

Localised drainage to intercept runoff and seepage from higher ground

Figure 9.6 *Where the field is affected by water from higher ground, interception drains can be effective. They could be —* top, *subsurface pipe where seepage is the problem or* bottom, *open ditches where runoff and/or seepage are a problem*

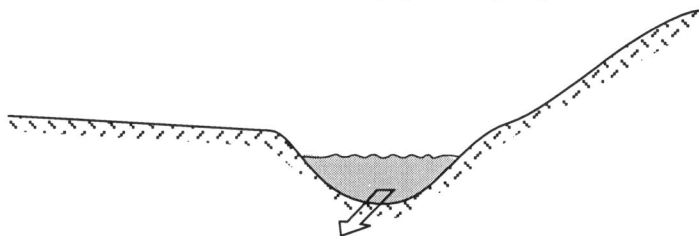

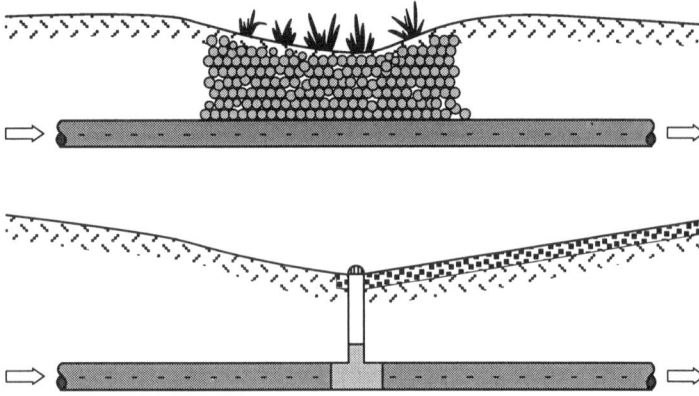

Figure 9.7 *Localised drainage spots –* top, *in the field or* bottom, *associated with paved areas and sheds can be incorporated easily*

The flow rate along the drain depends on its size and slope. The flow rate into them depends largely on the permeability of the soil they are installed in, which for a particular soil type, dictates the depth and spacing they should be installed.

Drain design is also dependant on the presence or absence of any impermeable layer which might be preventing vertical movement of water through the soil profile. If such a layer exists in reasonable proximity to the root zone, the drain is likely to require a higher capacity, and may be best located on the impermeable material.

Where a subsurface drain assists control of a rising watertable, they are sometimes referred to as "relief" drains.

Figure 9.8 *Various factors determine the design of a subsurface drainage system.*

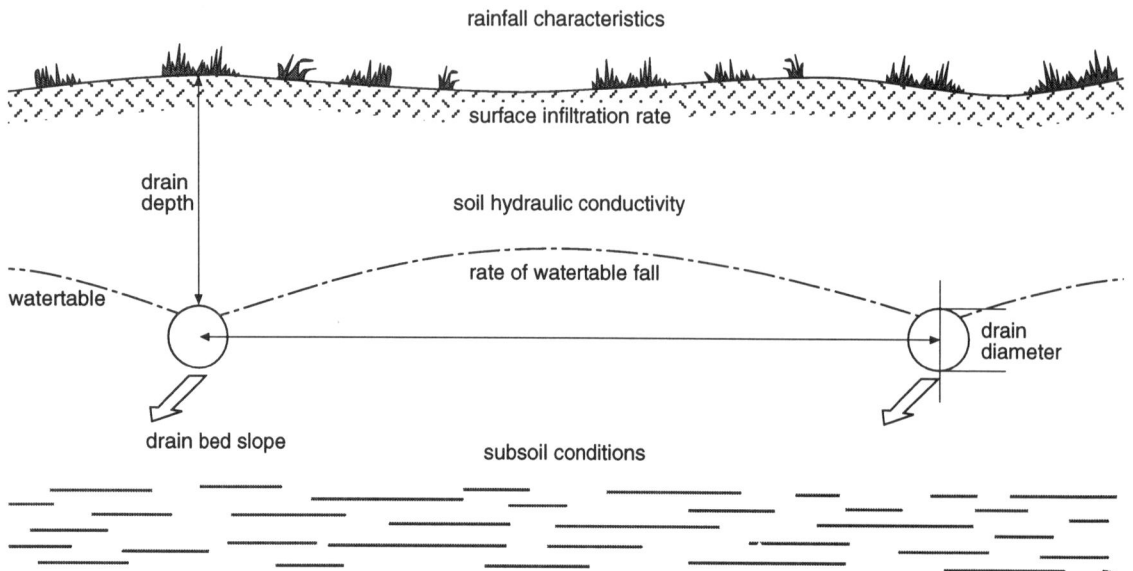

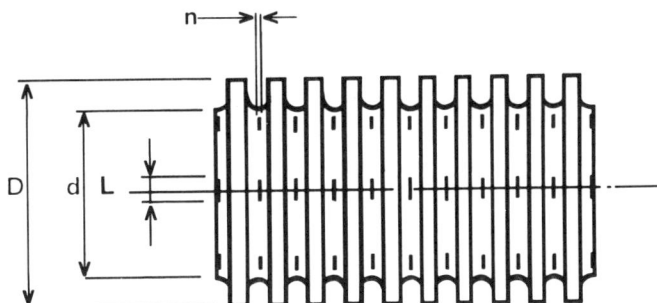

Class	Nominal Diameter (mm) D	Outside Diameter (mm) D	Inside Diameter (mm) d	Slot Size (mm) n x L	No. of Slotted Rows	Water Entrance Area (mm²/m)	Coil Length (metre)	Coil Weight (kg)
400	50	50	44	1 x 4	6	1800	20/200	3.2/32
400	65	65	55	1.1 x 5	6	1600	20/200	4.6/46
400	80	80	68	1 x 5	6	1600	20/100	7.0/35
400	100	100	86	1.1 x 7.4	6	2100	20/100	8.6/43
200	160	160	138	1 x 5	6	3000	20/60	16/48
200	200	200	175	1 x 5	6	300	25	24
1000	65	65	55	1.1 x 5	6	1600	100	31
1000	80	80	68	1 x 5	6	1600	100	48
1000	100	100	86	1.1 x 7.4	6	2100	100	67

MAXIMUM FLOW CAPACITY							
GRADIENT	FLOW AND VELOCITY	DN 50mm	DN 65mm	DN 80mm	DN 100	DN 160	DN 200
1:50	cubic m/h	2.5	4.6	8.3	15.8	57.7	110.6
	m/s	0.5	0.5	0.6	0.8	1.1	1.3
1:100	cubic m/h	1.8	3.3	5.8	11.1	40.5	77.7
	m/s	0.3	0.4	0.4	0.5	0.8	0.9
1:200	cubic m/h	1.2	2.3	4.1	7.8	28.4	54.5
	m/s	0.2	0.3	0.3	0.4	0.5	0.6
1:300	cubic m/h	1.0	1.9	3.3	6.3	23.1	44.3
	m/s	0.2	0.2	0.3	0.3	0.4	0.5
1:500	cubic m/h	0.8	1.4	2.6	4.9	17.8	34.2
	m/s	0.1	0.2	0.2	0.2	0.3	0.4
1:750	cubic m/h	0.6	1.2	2.1	4.0	14.5	27.8
	m/s	0.1	0.1	0.2	0.2	0.3	0.3
1:1000	cubic m/h	0.5	1.0	1.8	3.4	12.5	24.0
	m/s	0.1	0.1	0.1	0.2	0.2	0.3

Figure 9.9 *Product specifications for perforated polythene drainage pipe (Courtesy: Vinidex Tubemakers P/L)*

Subsurface drainage materials and installation

Traditionally, earthenware tiles were installed in trenches, end to end. Water enters the drain in the small crack between adjacent tiles. Because the base of the trench is constructed along a slight predetermined gradient, water entering the drain can flow along it. A filter cloth can be placed over each join to

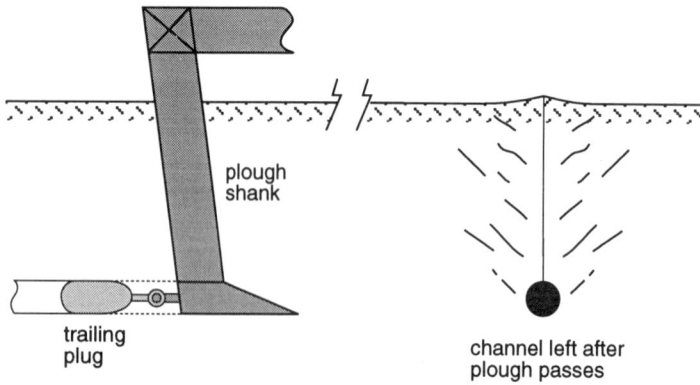

Figure 9.10 Top, *construction method for a mole drain. Bottom, mole plough*

help prevent earth being washed into the drain. Coarse gravel can be used to assist filtration, to increase the effective cross-sectional area of the drain, and possibly to provide greater stability to the trench surrounding the drain.

Perforated plastic pipe is now in widespread use. There are two main types; slotted PVC (used also as seepage pipe), and more commonly, perforated corrugated polyethylene pipe. Water enters the slots or perforations, which are manufactured around the full circumference and along the full length of the pipe, then flows along inside the pipe.

Plastic pipe material is relatively cheap, and easy to use. The polythene material is available in coils of long length, and is able to be installed in curved trenches. The corrugations impart strength to the pipe against crushing, whilst ensuring flexibility.

A filter "sock" is available to help prevent fine material washing into the pipe, which would encourage siltation downstream and possibly destabilise the trench. Gravel packing

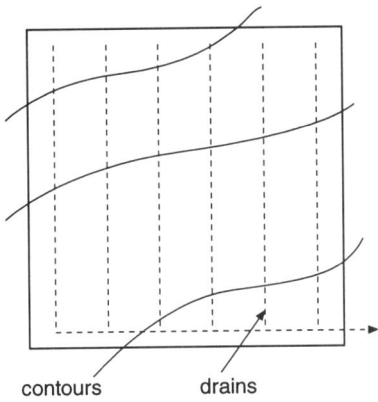

contours   drains

Figure 9.11 *Grid system*

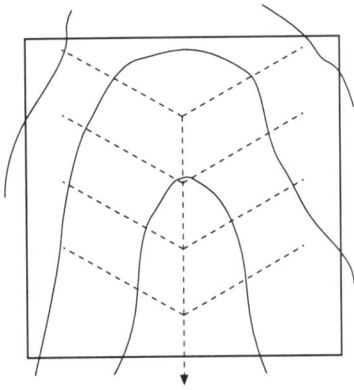

Figure 9.12 *Herringbone system*

Figure 9.13 *Random drainage directed to localised wet spots* – left, *schematic and* right, *actual.*

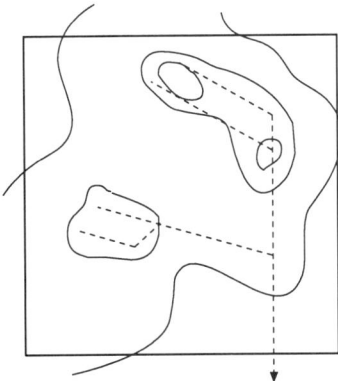

can also be used. Accurate trenching will ensure the correct pipe gradient. A trenching machine (such as a "ditch witch") may not provide accurate depth control (and hence gradient) and may be difficult to hand finish if too narrow. On the other hand, unnecessarily wide trenches should be avoided.

A mole drain is normally an unlined channel formed under the surface by the use of a mole plough. This resembles a heavy duty ripper, but it has a bullet head base to the tyne, with an oversize plug pulled behind. Sometimes, a low duty perforated plastic liner can be ripped in at the same time, formed from a coil of material anchored at one end.

The durability of an unlined mole drain depends largely on soil stability, and so they are generally considered temporary. Because their capacity is limited, more drains are required per hectare, but they are of lower cost to install. Their installation in unsuitable soil may cause the drain to collapse, or create an erosion tunnel.

Drainage systems

The layout of the drains needs to be matched to each particular situation. A number of options is possible:

1. A grid system may be necessary on flat or near-flat sites, where the value of the crop can justify the cost. Drains are located in a parallel grid, at a spacing determined by soil characteristics and the required drainage rate, and connected to an outlet pipe or ditch at one side (or in the centre for large fields).
2. Where the area being drained has a concave surface, or a narrow natural flow path, a herringbone system could be considered, especially where the drainage laterals are long.
3. Where drainage requirement is localised, a random drainage pattern is acceptable. These are not really

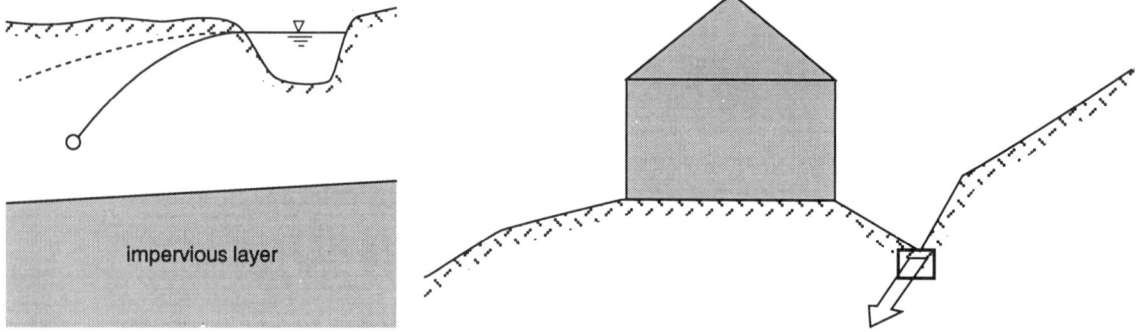

Figure 9.14 *Interception drainage from channels and buildings*

random, but directed to where they are needed, following natural drainage lines or connecting localised depressions to an outfall.

4. Interception drains are used to help prevent water reaching a particular area; for example:

   - Intercepting runoff from high ground.
   - Seepage control from groundwater or leakage from channels.
   - Around building sites.

5. Deep well drainage systems operate on a different principle. A well or bore is installed, and water extracted from it by pumping. This has the effect of lowering the watertable in a cone of depression around the well, possibly (and hopefully) extending for a substantial distance. This technique is usually applied on a regional basis, sometimes utilising multiple wells.

Drain design

The design of a drainage system must consider the rate at which excess water is to be removed, then the size and placement of the drains needed to remove it.

Drainage capacity

This refers to the peak volume or rate of removal of surplus water, and will depend on a number of factors:

   - Size of the drained area.
   - Permeability of the soil (its hydraulic conductivity, vertical and horizontal).
   - The rate at which water should be removed from the site, based on the tolerance of the crop to waterlogging, and the expected benefit to be derived.

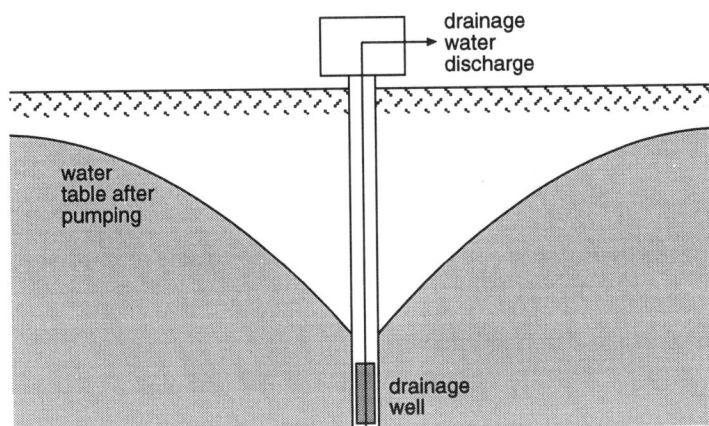

Figure 9.15 *Deep well drainage*

- The rainfall pattern of the area (or the irrigation practices employed).
- The runoff characteristics of the site.

For surface ditches, the potential for runoff from storm events will decide the capacity. In regard to subsurface drainage, the drainage capacity depends on how quickly water must be removed from the root zone. If, for example, the site is underlain by an impermeable layer, the rate that water moves through the soil to the drain will determine how quickly the watertable falls, for a given combination of drain depth and spacing.

Surface drain design

The relative elevations of the surface of the site and the discharge point will determine the fall available to the drain network, and therefore the bed slope. Frequently this will be a design limitation due to the topography of the site or a lack of appropriate discharge points. Where this is not the case, the maximum bed slope will be determined by the maximum allowable non-erosive water velocity (determined partly by the soil type), whether in the channel itself, or at structures installed within it (culverts, junctions, outlets). An overly flat gradient will require a larger channel cross-section to provide the necessary capacity, and will be susceptible to siltation.

In other respects, the design of a drainage channel is similar to that of an irrigation channel (discharge, freeboard, base width, shape and batter: see Figure 8.11). Structures in the channel should not restrict discharge.

The following general rules apply to drain locations:

- Follow the general direction of natural drainage lines, particularly with main drains, around which the drainage system, and property plan as a whole, could be based. Use the available gradient to best advantage.

- Avoid unstable and difficult soils.
- Drains could be located on property or field boundaries to minimise interference to farm operations. When located near trafficable headlands, ensure that accidents with machinery are avoided.
- Provide straight channels, with gradual changes of direction, and junctions that do not create turbulent flow.

Subsurface drainage design

Because of the high cost involved, and the complexities associated with predicting water movement through soil, expert advice should be sought for the detailed design of subsurface drainage systems. A number of theoretical and graphical techniques are in use to help determine the optimum combination of drain depth and spacing for any specific situation.

There is a definite relationship between depth and spacing; the deeper the drain, the further apart they can be placed for a given area and drainage capacity. Although deeper drains cost more to install, fewer of them are required. On very heavy soils, this guideline may not be entirely applicable, because of the very low hydraulic conductivity of these soils, a large number of relatively shallow drains may be required (and mole drainage becomes a more feasible option).

The drain depth is often limited by practical considerations; the limitations of the trenching equipment, elevation of the drain outlet, or the presence of an impermeable layer (where there is no advantage in going deeper). Consequently, the drain depth is often determined by site conditions, and the drain spacing is selected to match.

Research has also enabled other conclusions to be drawn (some of them are fairly obvious):

- Drains should be laid directly under areas where surface water accumulates.
- In the absence of impermeable layers, drain effectiveness increases nearly proportionately with depth.
- A gravel filter or envelope around the drain not only provides for sediment control, but also substantially increases the rate of flow of water into it from the soil. (Drain diameter is a less important factor.)
- Increasing the size or number of drain openings beyond a certain amount does not increase the effectiveness proportionally.

Apart from inadequate capacity and siltation, drains can fail by washing out. This is caused by too steep a gradient on

Figure 9.16 *Drainage water can be delivered to a disposal channel. A simple check valve will prevent entry of rodents and floodwaters. The last section of pipe should not be perforated.*

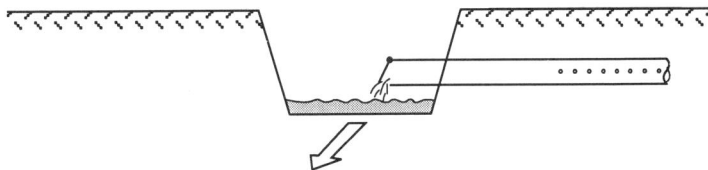

Figure 9.17 *On flat terrain or where deep drains prevent the use of channels, drainage water may need to be pumped from a sump.*

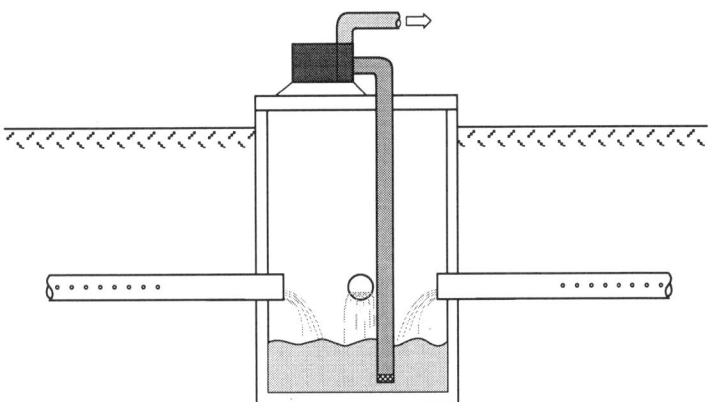

the drain, such that a full drain pressurises the downstream part of the drain. There is also a risk of water running down the outside of the pipe, although this risk should be minimised by correct installation. Sometimes the drain outlet is too small or becomes blocked or flooded.

Subsurface drains can discharge into a ditch for gravity disposal (Figure 9.16), or it may be necessary to discharge into a sump, and elevate the water to ground level by a pump (Figure 9.17).

Where an irrigation layout is designed with tailwater or runoff collection, it is desirable (but not always possible) to use a channel system that delivers water to a sump in close proximity to a water storage. Pumps can then be used to lift the water from the sump (the lowest part of the system) into the storage (the highest part of the system). This is easiest to achieve if considered as part of the whole property plan. (An example is shown in Figure 5.8.)

References

Awad, A. S., 1984. *Water Quality Assessment for Irrigation*, Advisory Bulletin No. 1, Department of Agriculture NSW.

Browne, R., 1984. *Irrigation Management of Cotton*, Agfact P5.3.2, Department of Agriculture, NSW.

Cornish, J. B., Murphey, J. P. and Fowler, C. A. (eds) 1990. *Irrigaiton for Profit: Water Force Victoria*, Irrigation Association of Australia, Numurkah.

Doneen, L. D., 1971. *Irrigation Practice and Water Management*, Irrigation and Drainage Paper 1. U.N.F.A.O., Rome.

Doorenbos, J. and Pruitt, W. A., 1975. *Guidelines for Predicting Crop Water Requirements*, Irrigation and Drainage Paper 24 U.N.F.A.O., Rome.

Garzoli, K., 1978. *Satisfying the Plants' Needs, Irrigation for the Orchardist and Vegetable Grower*, Agricultural Notes No. 6, Department of Agriculture, Victoria.

Kellerher, F., 1981. *Plant Water Use and its Estimation for Irrigation Planning*, in: Southorn, N. (ed.), *Proceedings*, Irrigation Efficiency Seminar, Hawkesbury Agricultural College, Richmond, NSW. October 1981.

Mitchell, P.D. and Goodwin, I., 1996. *Microirrigation of Vines and Fruit Trees*, Agriculture Victoria and Agmedia.

Murphy, J., 1991. *Efficient Use of Water* in: Energy Efficient Irrigation, State Electricity Commission of Victoria.

Prior, L. (ed.), 1993. *Drip Irrigation –A Grape growers Guide*, NSW Agriculture.

Somers, T. (ed.), 1996. *Grapevine Management Guide 1996–1997*, NSW Agriculture.

Southorn, N. 1995. *Farm Water Supplies – Planning and Installation*, Inkata, Sydney.

Standards Association of New Zealand,. *Code of Practice for the Design, Installation and Operation of Sprinkler Irrigation Systems*, NZS 5103: 1973.

Winter, E. J., 1974. *Water, Soil and the Plant*, McMillan Press, London.

Vinidex Tubemakers Pty Ltd, 1992. *Polyethylene Pipe Systems Technical Manual*.

Vinidex Tubemakers Pty Ltd, 1989. *The Water Supply Manual for PVC Pipe Systems*.

Index

Some other titles in the
PRACTICAL FARMING SERIES
Published by Inkata

BEEF CATTLE: Breeding, Feeding and Showing	Lucy Newham
CROP SPRAYING: Techniques and Equipment	Gary Alcorn
FARM AS A BUSINESS: Rural Property Planning	Jim Richardson
FARM BUILDINGS: Planning and Construction	Neil Southorn
FARM WATER SUPPLIES: Planning and Installation	Neil Southorn
FENCES AND GATES: Design and Construction	David East
FIRE FIGHTING: Management and Techniques	Frank Overton
LAND CARE: Rural Property Planning	Bill Matheson
PASTURE MANAGEMENT	Rick Bickford
PRUNING AND TRAINING FRUIT TREES	Warren Somerville
RISK MANAGEMENT: Rural Property Planning	Mike Krause
RURAL SAFETY: Stock, Machinery and General Hazards	Andrew Brown & Brian Lawler
RURAL SAFETY: Chemicals and Dangerous Substances	Andrew Brown, Brian Lawler & David Smith
SMALL PETROL ENGINES: Operation and Maintenance	Bruce Holt
SUSTAINABLE FARM ENTERPRISES: Rural Property Planning	Jim Richardson
WELDING: Techniques and Rural Practice	Peter Cryer & Jim Heather
WORKING DOGS: Training for Sheep and Cattle	Colin Seis